Explorations into a Dynamic Process-Oriented Soil Science

Explorations into a Dynamic Process-Oriented Soil Science

Douglas S. Frink, Ph.D.
Physical and Earth Sciences
Worcester State University
Worcester, MA

AMSTERDAM • BOSTON • HEIDELBERG • LONDON • NEW YORK • OXFORD
PARIS • SAN DIEGO • SAN FRANCISCO • SINGAPORE • SYDNEY • TOKYO

Elsevier
32 Jamestown Road London NW1 7BY
225 Wyman Street, Waltham, MA 02451, USA

First edition 2011

Copyright © 2011 Elsevier Inc. All rights reserved

No part of this publication may be reproduced or transmitted in any form or by any means, electronic or mechanical, including photocopying, recording, or any information storage and retrieval system, without permission in writing from the publisher. Details on how to seek permission, further information about the Publisher's permissions policies and our arrangement with organizations such as the Copyright Clearance Center and the Copyright Licensing Agency, can be found at our website: www.elsevier.com/permissions

This book and the individual contributions contained in it are protected under copyright by the Publisher (other than as may be noted herein).

Notices
Knowledge and best practice in this field are constantly changing. As new research and experience broaden our understanding, changes in research methods, professional practices, or medical treatment may become necessary.

Practitioners and researchers must always rely on their own experience and knowledge in evaluating and using any information, methods, compounds, or experiments described herein. In using such information or methods they should be mindful of their own safety and the safety of others, including parties for whom they have a professional responsibility.

To the fullest extent of the law, neither the Publisher nor the authors, contributors, or editors, assume any liability for any injury and/or damage to persons or property as a matter of products liability, negligence or otherwise, or from any use or operation of any methods, products, instructions, or ideas contained in the material herein.

British Library Cataloguing-in-Publication Data
A catalogue record for this book is available from the British Library

Library of Congress Cataloging-in-Publication Data
A catalog record for this book is available from the Library of Congress

ISBN: 978-0-323-16532-7

For information on all Elsevier publications
visit our website at www.elsevierdirect.com

This book has been manufactured using Print On Demand technology. Each copy is produced to order and is limited to black ink. The online version of this book will show color figures where appropriate.

Working together to grow libraries in developing countries

www.elsevier.com | www.bookaid.org | www.sabre.org

ELSEVIER   BOOK AID International   Sabre Foundation

Contents

	Preface	vii
1	**Introduction to a Dynamic Process-Oriented Soil Science**	1
2	**Soil Science in Historical and Philosophical Perspective**	5
	Soil Science in Historical Perspective	5
	Soil Science in Philosophical Perspective	10
3	**Self-Organization as a Result of Perturbations in a Dynamic Process-Oriented Soil Science**	13
	Kinds of Perturbations	14
	The Dynamics of Perturbations	17
	Soils as Self-Organizing Metabolic Systems	18
	Explanatory Soil Science	20
	Summary	22
4	**Soil Physiology: Processes of a Dynamic Process-Oriented Soil Science**	23
	Metabolic Processes: The Digestive System of Soil	23
	Respiration System	27
	Reproductive System	28
	Internal and External Communication System	29
	Biodynamic Feedback Systems	31
	Summary	33
5	**The OCR Procedure as Applied Dynamic Process-Oriented Soil Science**	35
	The Initial Iterations of the OCR Formula	35
	The OCR Procedure	35
	Evolving Applications of the OCR Formula	39
6	**Applications of OCR Dating**	41
	Geomorphic Studies	41
	Paso Otero 5, Argentina	41
	A Look at Landforms at Regional and Global Scales	47

    Unique Formation of Pimple (Mima) Mounds   53
    Misinterpreting Land Formational Processes   56
  Environmental Studies   59
    Climatic Complexity   59
    Paleoenvironmental Reconstructions from Anthropogenic Landforms   63
    Environmental Perturbations in the Dzban Burial Mounds   68
    Paleoclimatic Reconstruction from a Natural Landform   70
  Cultural Studies   75
    Assessment without Desecration of Sacred Spaces   75
    Examining Stratified Multicomponent Sites   77
  Summary   79

7   Implications and Potentials of a Process-Oriented Soil Science   83
  Carbon Sequestration and Climate Change   85
  Soil Health Indices   85
  Waste and Pollution Management   86
  Soil Geography   86
  Designing Sustainable Geo-ecosystems   87
  Agricultural Practices   87
  Geo-archaeology   88
  Astrobiology   88

8   Summary   91

Bibliography   95

Preface

One generally enters into scientific analysis of a subject with certain expectations based on one's learned or experienced history. Indeed, the very foundation of modern science is founded on the replication or reproducibility of results. Any new discovery or idea is validated by its reproducibility. This expectation of reproducibility is so ingrained in the scientist's mind that when confronted by unexpected results the first reaction is normally to question the analysis; what mistake did I make that led to these spurious data? When continually confronted by such unexpected results, the scientist will experience a crisis of faith resolvable only by entertaining the question that perhaps it is the expectation, and not the analysis, that is flawed. It is through such experienced crisis and resolution that new ideas emerge.

Questioning one's expectations is essentially a two-step process that begins with a deconstruction followed by a reconstruction of one's expectations aligned with an alternative or perhaps even a new paradigm. During the deconstructive phase one confronts various unquestioned assumptions, biases for and against certain ideas or ways of thinking. This is not to say that one need be judgmental toward these discovered biases, as they are neither good nor bad; they exist for various reasons intended or accidental, but in their existence they constrain one to a limited view of the phenomenon of inquiry. Those constraints are the very foundation of our expectations, and only by first removing such constraining biases are we able to reconstruct our expectations.

Reconstructing one's expectations along the lines of an alternative paradigm is an expedition into the unknown with few guideposts. It would be pure hubris to think that any new or alternative paradigm would be without its own set of biases constraining the scientific inquiry and thus the expectations of results. There are no *a priori* standards for judging the efficacy or detrimental possibilities of this set of new biases. These conditions are discovered only after the reconstruction is complete. Often the discovery is that one has selected a paradigm upon which a flimsy argument has been constructed, and thus one must admit to starting over again. In this process of reconstruction truth is relative; one must be resigned to selecting the one that works best.

This book is about my own experience as a scientist studying soils: my experience with unexpected data, my process of deconstruction and reconstruction of expectations. This book is the culmination of a study of soils that has extended over 25 years with analyses on over 7,000 samples from around the world, initially to understand why subsoil anthropogenic organic carbon behaves differently from surface-soil organic carbon. But the search to discern the reasons why this should be so has led to the development of new ways to think of, to characterize, and to understand soil, not as static geological detritus affected by organisms and weather, but as dynamic self-organizing systems that embody many of the fundamental processes usually associated only with biotic organisms. This new understanding of soil's complexity has potential significance for the sequestering of organic carbon in soil, the construction and use of soil health indices, policies and management of human-generated wastes and pollutants, agricultural practices, planning and development of sustainable geo-ecosystems, archaeological investigations, and even the search for extra terrestrial life.

A process such as this, taking place over the better part of my intellectual lifetime, would not have been possible without the help, guidance, critiques, and support of many people, too many to call out and thank individually. Over the years hundreds of colleagues supplied soils for analysis from many different contexts around the world. Dozens of coworkers over the years assisted in the analyses and edited the various papers and reports resulting over the years. A few individuals, however, deserve my special thanks. I am forever indebted to my two children, Ryon and Hilary, who refused to take my ideas at face value, requiring instead that I completely defended them as clearly as possible. I am most appreciative of Donald Johnson, professor emeritus at the University of Illinois, Urbana–Champaign, and Johannes Loubser, for their nearly two decades of support, and Professor Ronald Dorn, at Arizona State University, for over one decade of support and advice. Special thanks go to Dave Whitley, archaeologist and rock-art specialist, for his tutelage and advice in the writing of this book, and to Maximilian Baldia, archaeologist, for his review and critique on an earlier draft of this manuscript. Additionally, I am indebted to the many authors cited in the bibliography for the insightful ideas they have provided.

1 Introduction to a Dynamic Process-Oriented Soil Science

Farmers, gardeners, and fertilizer salespeople yearly submit samples to soil laboratories to determine the nutrient needs of their various crops. Soil science developed from, and for, this target audience. The theories and foundations of modern soil science are intentionally directed toward addressing the pragmatic needs of agriculture. As such, soil science has been primarily concerned with the identification and classification of soils, and their distribution across the landscape. Modern soil science, however, places little emphasis on explanatory models concerned with how soils work [1]. Although I did not recognize it at the time, soil tests I first conducted almost three decades ago illustrated this fact.

In 1982 I analyzed three soil samples obtained from a multicomponent archaeological site in Connecticut. I used the Ball Loss on Ignition [2] and the Walkley-Black Wet Oxidation [3,4] tests in my analyses. These are procedures commonly employed in USDA Extension Service and university soil labs to characterize soil organic carbon. They normally yield similar results for organically enriched surface-soil horizons (i.e., the A and E horizons, and the plow zone), with the Wet Oxidation method recovering 75% to 80% of the soil organic matter recovered by the Loss on Ignition method [5–7]. Measured results of both procedures are usually expressed in terms of amounts of organic carbon. These are then converted to organic matter, which is roughly 51% organic carbon for plow-zone soils [5]. My 1982 research area, formerly in agricultural use, had been abandoned to forest growth at the time of my sampling. Three archaeological hearths (prehistoric campfires) were recovered from below the plow zone in the Bs horizon, all at roughly the same depth below the surface. Artifacts recovered during the excavations indicated that these hearths dated from the Late Archaic [ca. 5,000 to 3,500 years before present (YBP)] to the Early Woodland (ca. 3,500 to 2,000 YBP) periods of Native American occupation.

Surprisingly, the two procedures did not yield comparable results: the organic matter contents of the three sub–plow-zone cultural features, as determined by the two different analytical methods, differed significantly. Yet the published soil science literature clearly showed a narrow range of variability between the two procedures, when applied to surface plow-zone soils [5–7]. The results from the Walkley-Black Wet Oxidation procedure for my three sub–plow-zone soil samples returned values 3 to 4 times less than the total organic carbon values found through the Ball Loss on Ignition procedure.

The literature at that time provided no clear explanation for the different results obtained from plow-zone versus sub–plow-zone soils, including whether these differences might be due to the effects of physical, chemical, biological, or anthropogenic processes. Furthermore, I had no idea how these differences might relate to the spatial distribution of soil at local to global scales [8–10].

This initial set of anomalous results was the spark for what has been a more-than-25-year-long study of soils, directed toward understanding why subsoil anthropogenic organic carbon (i.e., carbon from archaeological features) behaves differently than surface-soil organic carbon. This has required the analysis of over 7,000 samples obtained from natural and anthropogenic landforms from around the world, including tropical, temperate, subpolar, and desert regions. It has also resulted in the development of an index for characterizing soil organic carbon and its age, called the Oxidizable Carbon Ratio (OCR) procedure. Determining why standard organic carbon characterization procedures work differently in plow-zone and sub–plow-zone situations furthermore has led to the development of new ways to think of, to characterize, and to understand soils, not as static geological detritus affected by organisms and weather [1], but as dynamic self-organizing systems that embody many of the fundamental processes usually associated only with biotic organisms. Temporally dependent soil characteristics must be conceptualized dynamically, in a fashion that accommodates explanatory questions—"how?" and even "why?" [11,12]—even though soil as a physical object may be *described* in terms of where it is located.

Ideas, like things, have a history influencing how they develop and constraining what is essential to their composition. A scientific discipline like soil science develops in response to a societal concern or need. In the next chapter I begin by outlining the historical and philosophical roots of contemporary soil science, and how social and environmental conditions in Europe led to a science focused on identification and classification as the primary goal of the discipline. But contemporary societal concerns and soil science needs require more than an inventory of things to be identified and classified. Soils are dynamic, dissipative, and self-organizing systems whose study will require changes in how soils are perceived.

In Chapter 3 I discuss how modern soil science treats perturbations, followed by a characterization of perturbations in general. Perturbations are ubiquitous throughout time and space and represent a large source of both kinetic and potential energy for use by complex systems. The roles of perturbations in dynamic process-oriented systems are examined with specific reference to self-organizing systems, and the effects of perturbations at various scales on the structure and organization of these systems in accord with the Second Law of Thermodynamics. Soils, when viewed in terms of perturbations, exhibit the behaviors of a self-organizing system. Soils are energetic, concentrative, destructive, medium-forming, and transportive. These are the same terms Vladimir Vernadsky and others use to define the dynamic behavior of living systems.

Although operating at different spatial and temporal scales, biotic living systems and abiotic systems can be explained in similar *physiological* terms, as initially suggested by James Hutton, the father of modern earth sciences. I introduce the concepts of a soil metabolic system in Chapter 3, elaborating on and expanding this physiological explanation of soils to include respiratory, reproductive, information communication, and feedback systems in Chapter 4. In modern soils, biotic systems dominate in these processes. However, evidence of parallel abiotic processes remains in many of these physiological processes, suggesting a prebiotic template with the possible explanation for early biotic systems as an emergent property of prebiotic soils.

The next two chapters move from the theoretical to the practical. My attempts to understand and explain the apparent anomalous results obtained from the Walkley-Black and Ball Loss on Ignition procedures when applied to subsurface and archaeological soils resulted in the development of the OCR procedure, which combines these two organic carbon analyses to calculate age estimates for soil organic carbon. This procedure has also proven to be a valuable tool for the physiological assessment of soils as dynamic systems. I outline in Chapter 5 the development of the OCR procedure, the early false assumption made, and the insights these errors provided on how soils behave.

In Chapter 6 I use several case studies demonstrating the use of the OCR procedure in the physiological assessment of soils resulting in an explanatory approach to these soils. These case studies highlight the importance of distinguishing and sequencing the perturbational history of soil. From this perturbational history an understanding of when and under what conditions the soil's physiological processes effect the trajectory of pedogenic development.

In the final chapter I briefly touch on some of the many areas of study in need of a process-oriented soil science to overcome the limitations imposed by modern soil science. These areas of study fall mainly in the fields of physical geography, archaeology, and environmental studies, and also include aspects of meteorology, astrobiology, carbon sequestration, sustainability, and hazardous waste management.

First, it is important to understand what modern soil science is and how it came to be; its epistemological perspective and foundational paradigms. From this foundation I will construct in subsequent chapters the framework for a process-oriented soil science.

2 Soil Science in Historical and Philosophical Perspective

Archaeologists, paleontologists, geomorphologists, and many other kinds of scientists recognize the importance of studying the past, partly to appreciate the present. Although concern with the past is not an issue for all soil scientists, all of them likely appreciate that existing soil conditions and phenomena are the product of earlier processes and events. Equally, however, soil science itself is a historical product of earlier needs, interests, concerns, and philosophies. Understanding the status of soil science today, both its significant contributions and its potential shortcomings, is greatly enhanced by contextualizing its current intellectual agenda in terms of its own history. Here I will examine the interactions of political, climatic, economic, and demographic systems of European cultures from the Late Roman era through the 19th century, when modern soil science emerged as a discipline of study. This history is neither exhaustive nor complete. Rather, my intent is to link the ideas and focus of modern soil science to its historical roots.

Soil Science in Historical Perspective

Modern soil science's primary focus has been the identification and classification of the soils covering the Earth's surface. Soil science is one of the more recent physical sciences, emerging from its folk-based precursors only in the late 1800s. This occurred well after most of the physical earth sciences such as geology and geography had developed.

Soil science became necessary as a result of the interactions of political, climatic, cultural, and demographic systems. Its historical context can be traced back to the Late Roman period in Europe. Small privately held farms were abandoned due to high taxation by an indebted empire as a result of excessive war and corruption. These abandoned fields were acquired by wealthy urbanites as investment properties, a hedge against likely unstable political and commercial futures. Thus, large tracts of land were owned and controlled by a minority of the people following the collapse of the Roman social, economic, and political system after AD 476. The masses found themselves landless and without means of procuring food [13]. The feudal system was adopted at this time and was maintained throughout Europe well into recent history. This system was governed in part by the rules of relationship between the feudal lords and the masses, who returned to work the land as peasant farmers. These rules were established during the Late Roman period and were maintained by the only stable cultural–political system in the post-Roman era, the Catholic Church. These rules defined the rights and duties of the peasants in terms of their obligations to, and expectations from, the manor lord within a strict hierarchical social structure [14].

The production of grain, primarily wheat, constituted the primary focus of agriculture; it was sold at market to profit the manor lords, who then paid taxes to the fiefdom lords and the church. Small-scale animal husbandry and family-controlled kitchen gardens supplemented the needs of the peasantry. This system worked as long as environmental constraints allowed for a dependable yearly harvest [14].

Data from tree ring analysis [15], reconstructed sea surface temperatures [16], and residual atmospheric ^{14}C [17] provide proxy measures for both temperature and rainfall. These records indicate an ideal climate for grain production in Europe during this post-Roman period. Variations in climatic extremes were minimal, with temperatures above and rainfall below present conditions. This period is known climatically as the Medieval Warm Period (approximately AD 700–1300). Agricultural and manufacturing production rarely exceeded the demand.

The climate in Europe, and throughout most of the world, changed about AD 1300 [16]. The relatively persistent North Atlantic pressure cells (a couplet of Azores high-pressure and Greenland low-pressure cells) shifted to the south and west and increased their rate of oscillation compared to their earlier relatively stable positions. The reasons for this shift are not yet fully known, but it may have been the combined results of solar radiation variability and gravitational influences from the combined planets and the sun, or changes in the thermohaline circulation in the northern Atlantic Ocean [18], or some combination thereof. The effect on Europe's weather patterns was increased variability in climatic extremes, a general decrease in temperature, and a general increase in rainfall [19]. This period was the beginning of the broadly defined Little Ice Age. Weather patterns throughout this period were less predictable and included more frequent and larger storms in Europe, while extremes of cold, heat, and rainfall were more common. Continued reliance on wheat as the dominant staple, coupled with increasing populations, increased the risk of crop failure in the face of climatic uncertainty [20]. These less-favorable growing conditions for grain resulted in poorer yields and periodic crop failures, sometimes extending over several consecutive years. Combined with the marginal yields of subsistence farming under the feudal system, this ultimately led to widespread and often catastrophic famines throughout large areas of Europe.

Cultural adaptation to this high-risk environment became evident at this time. A low-technology agricultural revolution began in Flanders and the Netherlands as early as the 14th and 15th centuries. Farmers began experimenting with lay farming, including deliberately growing animal forage and cultivated grass for cattle. Animals prior to this time were fed in the wild or off gleaned fields after harvest. Farmers planted field peas, beans, and especially nitrogen-rich clover, all of which provided food for humans and beasts. Additionally, buckwheat, furze (gorse), and turnips were grown for feeding animals on land formerly allowed to lay fallow. The amount of fallow land contracted drastically until it was virtually nonexistent [19].

Land reclamation, primarily consisting of draining wetlands and coastal flood areas, began at about this same time. Concurrent with these changes was the emergence of profitable markets for expanded produce in Europe's growing cities. "These qualities of adaptability and opportunism came to the fore in sixteenth-to-eighteenth century Europe, when adequate food supplies in the face of changeable weather conditions became a pressing concern" [19; p. 103].

A revolution in thought accompanied this revolution in agriculture. Deteriorating weather conditions may have precipitated changes in agriculture, but the failure of feudalism under the political management of the Church was perceived as the underlying cause of food shortages and the resulting economic and political unrest. The Church was the dominant institution throughout Western and Central Europe during the medieval period, and it had set Aristotelian philosophy as its paradigm for intellectual pursuits [21].

The Age of Enlightenment, building on ideas developed during the Reformation of the 14th and 15th centuries, brought about changes in ideas of law and politics, as well as the substantial beginnings of governmental planning [21]. The goals of science shifted to the discovery and use of the wider world's resources, as evident in the following quote from Francis Bacon: "let the human race only recover its God-given right over Nature, and be given the necessary power; then right reason and sound religion will govern the exercise of it" [22; Aphorism 129, p. 131].

This new Cartesian philosophy emphasized three primary concepts: reason, nature, and progress [23]. Institutional and cultural conditions, in enlightened eyes, had corrupted the natural working of reason in most human beings. The rules governing the relationship between the peasants and the feudal lords had become debased and nonfunctioning. The rise in the power of secular kings had eclipsed the social order and power of the church. Reason, properly working, enabled human beings to discover, or rediscover, nature beneath the concealing corruptions of religion, social structure, and convention. Nature was equated with the good and the moral, while the burdens of irrational social customs and traditions, accumulating historically, were "un-natural." The correct (i.e., with reason) study of nature provided the foundation and means for deconstructing the Church's authority over secular institutions, and the feudalistic system itself became the "un-natural" cause of strife [23].

A new organizing concept of thought emerged in opposition to the deemed "un-natural" Judeo-Christian tradition embodied in the fall from grace and in opposition to the various Hellenistic cyclical theories. This new paradigm was progress. Progress was linear and necessary and all events and phenomena could be understood empirically through hierarchical "chains of causality." Chains of causality codify the linear cause-and-effect relationships between things resulting in perceived events and phenomenon. The concept of progress set the stage for the linear theories of evolution as later proposed by Charles Darwin and Herbert Spencer [23].

Coupled with this overriding concept of progress was a change in perception of the human relationship to nature. John Locke argued that the previous condition of humankind under feudalism was that of *animal laborans*. People were relegated to the cyclical task of endless labor for consumption as a biological necessity. Progress could be enacted only by *Homo faber*, man as the fabricator, or maker, of his world. *Animal laborans* nourished life through his physical efforts, and through domestication established himself as lord and master over plants and animals. *Homo faber*, however, established himself as lord and master over the whole earth [24]. All things from this perspective were reduced to simple use objects and means to a human-defined end.

Natural progress governed by reason became the perceived answer to the human food crisis during the Little Ice Age. Extensive research and writing on agriculture appeared throughout Europe at this time. For example, some 250 works on agriculture were written in France alone in the 16th century, most about ways to increase and diversify production. Some of these studies proposed soil classification systems and methods for their treatment, while others focused on the development of new crops like turnips, rice, cotton, and sugarcane [19]. Similar activity can be found in the histories of most other European countries; for example, in Prussia under Frederick the Great and his son, Frederick the Second [25], as well as in the newly founded and agriculturally floundering early-19th-century American republic [26].

The European cultural adaptation to the food crisis that resulted from increased climatic variability was the intensification of agricultural production by way of innovations in agriculture, promoted under the concept of progress. Modern soil science emerged in the late 19th century from various localized folk-based movements addressing social concerns about the frequent and persistent famines plaguing Europe throughout the 17th, 18th, and 19th centuries. Vasily Dokuchaev, a Russian geographer, was the first to recognize that geographical variations between soils could be explained by climate, topography, and time, as well as biology and geology, or parent material [27]. Dokuchaev [28], the father of modern soil science, "writing of his first classification, stressed 'the close connection existing between soil types on the one hand and yields, kinds of crops, and husbandry methods on the other'" [29; p. 15]. Thus, to improve agricultural production, certain soils, such as Chernozem formed in loess deposits, should be selected for crops such as wheat. Soil classification and mapping became the predominant pursuit of modern European and American soil science well into the middle of the 20th century.

Early soil classification systems were not uniform and tended to reflect national and regional concerns. Unification and improvements to these various soil classification systems came about with the enumeration of the five-factor model by Hans Jenny [30] in 1941. Building on Dokuchaev's initial studies, Jenny developed a factorial model explaining soil genesis by providing a mathematical basis for classifying soils. The five factors enumerated by Jenny were climate, organisms, relief, parent material, and time (CLORPT) (Figure 2.1).

Although differences still exist between the soil classification systems of Russia, England, and the USA, for example, Jenny's five-factor model allowed for easy translation and comparisons between similar soils. Mollisols found in Peoria, Illinois, and Chernozem found in Kiev, Ukraine, could now be linked through similarities in climate, relief, parent material, and time, and differences resulting from different organisms could be deduced.

$$s = f(cl, o, r, p, t)$$

Figure 2.1 Jenny's five factors of soil formation formula.

Adoption of the five-factor paradigm lead to the establishment of the various chains of causality, the sets of initial conditions and processes that would result in the classified soils. Soils are traditionally described and classified according to the kinds and degrees of development of horizons that result from 13 paired processes (Figure 2.2) [31]. Each of these 13 paired processes defines the chain of causality and describes environmental conditions leading to soil horizon development and the countering effects of perturbations. Most soil studies have focused on these processes leading to soil horizon development, with only a few notable exceptions [32,33,34] providing insights into perturbations.

The model [28], paradigm [30], and chain of causality [31] of modern soil science successfully addressed these 17th-, 18th-, and 19th-century social concerns on crop production with the launching of the "green revolution" during the latter half of the 20th century [35]. Initially, Dokuchaev's conceptualization of the five-factor model was explicitly dynamic and process oriented [36]. However, the specific direction of soil science as it was transmitted and evolved, particularly in the USA, became increasingly static and focused on the enumeration of soil geochemical constituents [37].

The choice of paradigms and epistemological perspectives predetermines what variables and phenomena are possible for study. This fact is epitomized in the story of the black swan. Prior to the discovery of the black swan in Australia, Western science maintained that all swans were white, a position so entrenched that it was commonly used in logical syllogisms. This firm belief resulted in the misidentification of this Australian bird and delayed acceptance of the black swan within its appropriate classification [38].

1. **Eluviation and Illuviation**, the movement of material out of or into different soil horizons,
2. **Leaching and Enrichment**, the movement of material out of or into the soil,
3. **Erosion and Cumulization**, the removal or addition of material to the soil surface,
4. **Decalcification and Calcification**, removal or accumulation of calcium carbonates,
5. **Salinization and Desalinization**, the accumulation or removal of soluble salts,
6. **Lessivage and Pedoperturbation**, the mechanical migration of clays downward through the profile or the physical homogenization of soil textures,
7. **Podzolization and Desilification**, migration of sesquioxides or silica out of the upper portions of the soil profile,
8. **Decomposition and Synthesis**, the breakdown or formation of organic material,
9. **Melanization and Leucinization**, the darkening or paling of soil horizons,
10. **Littering and Humification**, the accumulation of organic material on the soil surface and its transformation into humus,
11. **Ripening and Mineralization**, the chemical, biological and physical changes evident in aerated waterlogged organics and the release of oxide solids through decomposition of organic material,
12. **Rubrification and Gleization**, measuring the degree of sesquioxide oxidation or reduction, and
13. **Loosening and Hardening**, the increase or decrease voids in the soil.

Figure 2.2 The 13 paired processes.

Soils from the start were never the primary focus of soil science; plants were. Soil and soil science became the means to obtain the human ends of more and better food plants. It was thus inevitable that soils were, on the one hand, the cause of good or bad plant production, but also the result of the same variables that affect plant production: climate, organisms, relief, bedrock (raw mineral source), and time. This conflation of soils and plants was extended to include humans by the German chemist Justus von Liebig, who became known as the father of chemical agriculture as well as the father of modern nutritional science. Liebig reduced soil to a short list of essential nutrients required by plants for growth. Most important among these were nitrogen, phosphorus, and potassium, the N, P, K ingredients in common commercial fertilizers widely available today. In a similar reduction, his prescription for human nutrition consisted of a short list of macronutrients, chief among them being protein, which contains nitrogen, and carbohydrates and sugars, which contain phosphorous [39,40]. The use of potash as a seasoning salt provided potassium to the human diet [41]. The conflation of human and plant nutritional requirements represents a classic case of reductionist thinking that unfortunately still influences human nutrition today [40]. Thus, modern soil science as it has evolved is primarily concerned with the description and classification of soils rather than with explanations for how soil systems work [1].

Despite this narrow research focus, modern soil science remains foundational to the study of soil within many other disciplinary contexts. Jenny's [30] five-factor model of soil formation and Retallack's [1] universal definition of soil predominate within the discipline of soil science. This model and definition are the dominant and historically the only conceptual models governing soil science, and are taken to provide a sufficient explanation for all pedological and soil physical geographical phenomena. Pedagogically they are the first, or one of the first, concepts introduced to the student of soil science, physical geography, archaeology, and ecology. Furthermore, they inform all research concerned with soil genesis and classification in these various disciplines.

Soil Science in Philosophical Perspective

The structure of modern soil science proved itself more than adequate to address the 19th- and 20th-century agricultural and engineering concerns of Western civilization. The questions asked of soil by a global society at the beginning of the 21st century, however, have changed [42]. A soil science limited to description and classification cannot adequately address issues of sustainability, preservation, pollution, climate change and carbon sequestration, urbanization, and cosmology, as well as the ongoing demands placed on agriculture to feed a growing global population and on engineering for environmental hazard assessments and mitigations.

How a question is framed is an initial condition of science. If the question asks what something is made of or how many different kinds of a phenomenon exist, then the answer will necessarily involve description, classification, and/or simple quantification. If the question is how something works, then the answer will necessarily involve mechanics (chemical or physical). Addressing today's global issues and concerns as they pertain to soils will require questions framed in a process-oriented manner: how does a soil respond to specific perturbations (human-induced as well as natural), and what are the ramifications of its response in terms of feedback effects? Essentially the question becomes one of how a soil behaves as a system.

Modern soil science is a product of centuries of evolving philosophies based on the goal of understanding the whole in terms of its parts, in this case in terms of the 13 paired processes. Clarity in understanding the makeup, and thus the potential behavior, of the whole is obtained by refining the elementary parts. Science conducted under Cartesian philosophies has allowed for managed inventories, classifications, and experiments that relate specific causes to specific results among or between individual components. This approach to science, however, has also limited our ability to understand many of the behaviors of complex and dynamic systems.

This Cartesian approach tends to imbue components with a quality of static permanence. Understanding systems according to their parts requires that the individual components be isolated out of context and evaluated according to the linear effects they might have on other components. Such linear equations are fundamental to the Cartesian foundations of science. The system is understood according to the mechanics of its operational parts; unless a causal agent external to the component should act upon and thereby effect change to that component, it may be thought of as invariant [43]. This is the fundamental premise of modern science that the object of study must be perceived of and studied as being necessarily a system closed to external causal agents.

Complex and dynamic systems have characteristics and behaviors that cannot be reduced to an inventory of mechanized cause-and-effect equations between their components. They must be analyzed as whole systems. The relationships between components are not necessarily linear, and cannot be characterized simply through the enumeration of the parts. Dynamic systems are open to the exchange of energy, matter, and information with their environment, but their organization is closed. Their organization is defined and created by the nonlinear dynamic relationships (the systems structure) that exist between the components as a whole. They may be described as synergistic in that the sum of the various parts do not, in fact cannot, provide an understanding of the whole either as an entity or in terms of its behavior. This was amply demonstrated by Willstätter [44], whose research on enzymes in the early 20th century showed that once he had reduced an enzyme into its constituent components, it no longer functioned as a catalyst. The behavior of dynamic self-organizing systems is evident in the interactive structure comprising the relationships that exist between its components, and not simply the components themselves.

There are other ways of looking at the world that do not require the kind of reductionism evident in Cartesian science. Scientific pluralism suggests that there are many scientific methodologies accommodating meaningful inquiry, and multiple methods are necessary in the pursuit of scientific knowledge. This idea is firmly premised on the fact that the real world is complex, requiring that we ask many different kinds of questions of it. Different ways of thinking are required to answer these different kinds of questions. As Midgley [45; p. 50] notes with respect to multiple lines of inquiry: "The relation between them is not linear but convergent."

The five-factor model and the chains of causality defined by the 13 paired processes of modern soil science are effective for simplifying the complexity of one aspect of food production: soil as a plant growth medium. But they are not particularly helpful in addressing questions about soil as a dynamic, dissipative, and self-organizing system. They cannot show us the essence, or "soilness," of soil, or offer meaningful explanations for how soils behave. Reducing soil to its basic building blocks forces one to ignore the dynamic relationships that exist between those building blocks. Yet it is these relationships, as much as if not more than the building blocks themselves, that differentiate soil from inert geological sediments.

Cartesian philosophy is not, nor has it ever been, the only philosophy to inform scientific research in the modern era [46]. Organicism that is informed more by inductive reasoning has offered an alternative, more holistic process-oriented approach to inquiry since the inception of modern Western philosophy [47]. Cartesian philosophy draws exclusively on classical Western thought, but organicism draws on both classical Western and Eastern (primarily Chinese) thought. The *Ta Chuan* (Great Treatise) of the *I Ching* (Book of Changes), for example, provided not only the basis of Leibniz's system of binary numbers but also the necessity for including dynamics in the assessment of observed phenomena [48,49].

Organicism as a valid scientific approach reaches its zenith in the metaphysics of Alfred North Whitehead [50,51] and in the process philosophies of the American Pragmatists, principally Charles Sanders Peirce and John Dewey [52,53]. The postmodern critique of the limits of Cartesian science is very much a continuation of organicism [45,54,55,56].

Process-oriented philosophy informing science has run concurrent to but in general remains unintegrated with Cartesian philosophy. The integration of these two philosophical traditions for soil science does not require the total overhaul of the discipline. Instead, it requires only a shift of focus from an inventory of constituent parts to an assessment of the dynamic relationships existing between these parts. A process-oriented soil science must start with an understanding of soil constituents as "things" but then add the process behaviors resulting from the dynamic relationships that exist between those constituents, and between the soil system and its environment. Key to this change in focus is an emphasis on perturbations, not as a force simply countering soil developmental processes, but as endemic and necessary to all soil development.

My approach in the remainder of this volume reflects and is intended to accommodate these points. While traditional soil science has proven extremely valuable, its historical origin in the practical realities of agriculture has limited it, theoretically, methodologically, and philosophically. A process-oriented soil science, in contrast, accepting a dynamic universe of perturbations and sensitive to the behaviors of self-organizing systems, is better suited to answering questions pertinent to research and to policymaking in several significant areas of global concern.

The next chapter begins with a discussion on how modern soil science treats perturbations, followed by a characterization of perturbations in general. Throughout the chapter the roles of perturbations in dynamic process-oriented systems are examined with specific reference to self-organizing systems.

3 Self-Organization as a Result of Perturbations in a Dynamic Process-Oriented Soil Science

Soil development has traditionally been viewed as resulting from two opposing processes: horizonation, the tendency to differentiate into separate horizons, and haploidization, the tendency for perturbations to homogenize soil [57]. Most pedological research has focused on the processes leading to horizonation, although there are some noted exceptions, such as Hole [32], Johnson et al. [33], and Wood and Johnson [34]. Together, these last authors present 10 pedoperturbational processes (i.e., processes that alter the physical and/or chemical properties of soil) to classify observed agents of haploidization. An 11th process, Anthroperturbation, a subset of Faunalperturbation, is commonly used by archaeologists to distinguish potentially significant cultural events that have altered either the soil or the context of archaeological artifacts or features (Figure 3.1).

Hole [32] earlier introduced the concepts of propedisotropism and propedanisotropism to describe variations of soil-mixing processes resulting from perturbations common to all soils. Some soil-mixing processes, propedanisotropism, lead to the formation of horizons through the process of horizonation, while other soil-mixing processes, propedisotropism, lead to the destruction of horizons through the process of haploidization. Unlike the concepts of horizonation and haploidization, which describe diametrically opposed processes, Hole's terms suggest states or conditions resulting from universal perturbations affecting all soils. Any given perturbation may appear to lead toward horizonation yet simultaneously lead toward haploidization as well, depending on observer bias, context, and scale of observed phenomenon [32]. Unfortunately, this idea never caught on in modern soil science.

Perturbations, then, need not be thought of as exclusively destructive. Rather, they may be considered a necessary aspect of soil development contributing to the organizational stability of the soil system. Soils when seen in this light have a history of recurrent perturbations consisting of those kinds (propedanisotropism) that lead to metastability or preservation of the system.

A soil may be viewed at two different levels of abstraction. A soil as a specific entity, such as Adams series (Spodisol), is described by its organization, the inventory of its constituent attributes organized into pedogenic horizons. At a different level of abstraction a soil's structure (the dynamic relationships between its constituent attributes) is constantly changing in response to perturbations. Textural sorting, weathering and clay crystal formation, illuviation of sesquioxides and clays,

Process	Soil-mixing vectors
Aeroperturbation	Gas, air, wind
Aquaperturbation	Water
Argilliperturbation	Swelling and shrinking of clays
Cryoperturbation	Freezing and thawing
Crystalperturbation	Growth and wasting of salts
Faunalperturbation	Animals (burrowers especially)
Floralperturbation	Plant (treefall, root growth)
Graviperturbation	Mass wasting (e.g., creep, solifluctation)
Impacperturbation	Comets, meteorites
Seismiperturbation	Earthquakes
Anthroperturbation	Human modified soil

Figure 3.1 Traditional classification of pedoperturbational processes.

and mineralization and sequestering of organic carbon are all examples of a soil's structure at a given time. However, these must be viewed as processes maintained or changed as a result of perturbations, and not confused with the soil's constituent attributes. Thus, preservation of the soil's organization as a specific entity through structural change triggered by recurrent perturbations results in a condition of structural coupling between a soil and its environment [58].

An individual soil, then, can be conceptualized as evolving from initial conditions along a unique trajectory defined by its organization and its relationship with its environment, through recurrent perturbations. Everything that occurs to the soil between its initial conditions and its present observed state constitutes the pedogenic history of that individual soil.

Generally the environment is thought of as effecting change to the soil through perturbations, but such unidirectional deterministic processes could not result in the metastable conditions of an organized system. Such one-way influence would lead only to the destabilization and ultimate destruction of the system. Instead, structural adjustments are made by the soil system in response to environmental perturbations. These triggered responses or adaptations in structure preserve the organization of the soil as a unique entity. In this way soil systems display self-organizing characteristics; indeed, they are complex self-organizing system affected by perturbations. Furthermore, as will become evident in the following section, as the self-organizing system adapts to a particular perturbation, it influences the nature of that perturbation; the dynamics are multidirectional.

Kinds of Perturbations

Perturbations are ubiquitous at all spatial scales and come in many different sizes and frequencies. They may, however, be thought of and treated collectively. The ubiquitous nature of perturbations is manifested as physical and/or chemical changes within the three dimensions of a soil. This demands that soils not be described just in terms of a two-dimensional profile, or even a three-dimensional pedon (a volume of soil having a hexagonal surface area with sides measuring 1 m). A consideration of perturbations requires that soil position, or development, ideally involves the sum of all perturbations that have occurred and the organizational responses of the soil system to these perturbations [59–62]. In these terms, a soil description is necessarily four-dimensional in the sense of requiring spatial and temporal data. Along with the three observable dimensions of height, width, and depth, a consideration of perturbations provides a measure of disturbance, displacement, and reorganization of the specific soil system. In essence, the speed and trajectory of soil components are being measured.

Perturbations in general can be characterized by certain structure and order. The size of a perturbation, relative to the system experiencing it, is inversely related to its frequency, and the degree and kind of perturbation defines the stability or instability of the system as a definable unit [33,63–65]. The gentle swaying of a tree in the wind translates to a nearly constant but infinitesimally small perturbation to the soil, whereas the infrequent impact of an asteroid would severely affect a soil, to say the least. Successful systems tend to be organized to absorb the plethora of constant small lower-order perturbations but do not overly invest in preparation for the infrequent few large higher-order perturbations. Thus, recurrent perturbations that are structurally coupled with the environment and that support the organization of the soil as a system are of two kinds: low-order perturbations that lead to the development of horizonation; and higher-order perturbations, contributing to haploidization. Higher-order perturbations are generally perceived to interrupt pedogenesis in traditional modern soil science. It is not that simple, however, as explained below.

If soil acts as a self-organizing system, then it must be homeostatic, if not homeorrhetic, maintaining its own organization constant [66]. Systems having set points between which they oscillate, such as a thermostat, are characterized as homeostatic, while systems having operating points defined by the system's organization are more appropriately referred to as being homeorrhetic [67]. Such systems depend on small perturbations, which we might think of as first-order perturbations, to supply energy for their various organizational processes [68,69]. Lower-order perturbations, then, are ubiquitous and constitute a constant component of the environment. Some higher-order perturbations are less often encountered but stronger in effect and challenge the homeorrhesis of a soil body. They may include mechanical events such as mass wasting [70], riverine erosion and deposition [71], solidification from seismic events [72], and biological events such as tree-tips [73], rodent activities [74–78], and human actions [79–83]. The net effect on the soil is a structural change in response to such higher-order perturbations. Soils as self-organizing systems are expected to maintain the trajectory of their pedogenic processes when the perturbations are minor or moderate.

These and other higher-order perturbations, which we might think of as second-order perturbations relative to the smaller first-order perturbations discussed earlier, such as human land use changes [84–89], fire [90–94], and environmental events on the order of decadal scale droughts, affect only the structure of the soil, leaving the essential organization that defines it intact. These structural changes may be invisible as the soil adapts to the perturbation, conserving its unique organization. The effects of these perturbations, however, are evident and measurable in the soil as structural changes [95,96].

Catastrophic or third-order perturbations are those higher-order perturbations that are expected to result in the complete organizational change of the dynamic pedogenic system [97]. If the organization of a system changes, then its identity changes and it becomes another system [66]. For example, when a soil is deeply buried, isolated from essential environmental perturbations, under pressure, and heated, it becomes rock (lithified). Alternatively, for soils removed by glacial actions and subsequently redeposited, this change in identity would appear as the commencement of a new soil whose components include, in whole or part, the components of former soils but with markedly different structural relationships.

Considering perturbations along with the three observable dimensions of height, width, and depth also allows a means for understanding emergent properties (*categorial novum*): those unexpected observations or unforeseen patterns that require new explanations [98]. Macklem [99] lists three requirements for emergent phenomena. First, they must be thermodynamically open systems; second, feedback loops must exist leading to nonlinear behavior; and third, component parts must be interconnected. Soils are thermodynamically open systems. Self-maintenance in the face of second-order perturbations is a nonlinear behavior—adaptive structural change. Soil systems are organized as interconnected components each linked to others through their metabolic processes.

New systems are manifest in changes in the system properties, such as self-organization [33,66], feedback processes [100,101], equifinality [102], and entropy analysis [60]. These new properties emerge, not *ad nihilo*, but out of the structural coupling of the system within its second-order-perturbation–filled environment. Thus, it is not the case that perturbations cause soil changes, but rather that perturbations allow emergent properties as new patterns in the relationships between components. Once a system has structurally adapted to the particular second-order perturbation, the recurrence of that particular perturbation becomes effectively a first-order perturbation relative to the structurally adapted system. As the self-organizing system adapts to a particular perturbation, it influences the nature of that perturbation. The interface between the system and its environment in this sense forms what Foucault [54] terms a "surface of emergence."

The effects of perturbations are also order-dependent. For example, common sense suggests that the effect of a rainstorm on a given soil will be different if it is preceded by a lightning strike in a forest rather than if rain precedes the lightning strike. In the first case one expects a lightning strike to cause a forest fire and the subsequent rain to result in severe erosion of the denuded soils. In the second the forest is wet from the rain and the lightning strike might affect only one tree, with no forest fire resulting.

It is important at this point to introduce a new term that I will use later in describing various case studies: *soil packages*. The punctuated dynamics of soil formation resulting from second-order perturbations may be thought of as resulting in separate packages of soil deposits. Normally, soils are described by their pedogenic horizons (A, E, Bs, Bt, etc.). However, soils may be described by what may be thought of as "event horizon." Packages of soils are related by depositional events and subsequent pedogenic restructuring processes. Clearly buried soils do have a distinct nomenclature based on their visible characteristics as indicated by such designations as 2Btb horizon. However, soil packages are different: they reflect dynamic processes that take place within a specific space and time but may not be visible within the framework of a single set of horizons in a traditional soil description. Second-order perturbations, then, are evident in spatial "packages" of time-dependent pedogenically active soils. Each package is bounded by the perturbation events that define its beginning and distinguish it from the end of the previous package. These defining events do not destroy the organization of the soil as a whole. Rather, they are another kind of structural adaptation that preserves the soil's organization as a system. This is evident in the immediate resumption of the same pedogenic processes at the perturbated soil's surface, initiating a new sequence of pedogenic activity. But despite the perturbational event, this new domain of pedogenic activity maintains the same soil taxum characteristics. This is similar to the annual growth rings in a tree; tissue once in active growth now dead, but still functioning as an integral part of the tree. The former domain of pedogenic activity is transformed to a component serving other functions in the system—a structural adaptation. Similar maintenance of soil organization in perturbated soils has been noted elsewhere in landslip erosion topo–chrono sequences [70].

The Dynamics of Perturbations

But where do all these perturbations originate? Some perturbations result from other perturbations as consequences of dynamic system adaptations. Large third-order perturbations can be the source of a series of second-order perturbations that in turn result in the proliferation of first-order perturbations within the environment. For example, a tornado running through a forest is catastrophic to the trees that are uprooted, but the consequent open space enables increased opportunity for the seed germination of other trees, which in turn feed back on the whole forest system. Conversely, small first-order perturbations can combine to create second-order perturbations that in turn contribute to third-order perturbations. Lorenz's "butterfly effect" [103] serves as an example, where the flapping of the butterfly's wings might set in motion a series of events that culminates in the tornado that ruins the forest.

In a recent essay Dorion Sagan and Eric Schneider [104] suggest that the dynamics of change are the thermodynamic breakdown of existing chemical, physical, and thermal gradients. In this sense, a perturbation is the event of gradient breakdown.

Perturbations potentially represent a large source of both kinetic and potential energy that self-organizing systems use for their organization and dissipate back to their environment as entropy [105]. Ultimately, perturbations are the result of the one force in the universe that flows uphill thermodynamically: gravity [106]. The Second Law of Thermodynamics explains the tendency for all things to trend toward disorder and ultimately to rest in equilibrium in a completely entropic state. Gravity, on the other hand, can be thought of as playing a role countering the Second Law of Thermodynamics by bringing things together, creating gradients, and providing order. Both are ongoing processes that establish fundamental dynamics at a universal scale [107].

Considering perturbations as an essential component of soil provides considerable potential for a convergence of Cartesian and process-oriented science. Soils are not static geological detritus, but are in constant change. Perturbations issuing from a wide range of natural and human activities and from climate change are ubiquitous for all soils. Each individual soil has evolved the way it has as a result of its unique history of perturbations. As such, an understanding of soils and their genesis must include more than a simple classification of perturbations.

Attempts to deductively determine cause-and-effect relationships between specific perturbations and resulting soils are problematic because the temporal and theoretical constraints on experimental procedures prevent such inferences. For example, if a scientist were to design an experiment to determine if perturbation "A" results in soil "B," the scientist might consider subjecting a new or incipient soil to the perturbation and then observe its effect on the soil's development. But the rates at which soils develop, and the time depth required for the effects of that perturbation to become fully evident as a soil characteristic, operate at timescales of hundreds to thousands of years. Alternatively, a mature soil could be examined to determine the specific perturbation that caused a particular soil characteristic. The trouble with this approach is that the scientist would not necessarily know when or if a certain perturbation occurred in the past. More of a problem is the fact that such an experiment could not be limited to one single perturbation: perturbations are ubiquitous throughout time as well as space.

Inductively constructed theoretical models that link certain individual or groups of perturbations that lead to observable soil characteristics, however, can be constructed. Such models can then be tested against empirical data obtained from analyses of a wide range of individual soils.

Soils as Self-Organizing Metabolic Systems

Consideration of a soil's spatial and temporal context is a prerequisite to its sampling for analysis. This consideration must extend to the formational processes resulting in the feature or landform itself, as well as to the self-organizing formational processes of soil. Self-organizing systems are dissipative systems. They increase order within themselves by dissipating entropy back into their environment [105,108]. These systems are closed organizationally and independent of their environment. Self-organizing systems, however, are open to the exchange of information, energy, and matter with their environment [66], and through perturbations are coupled with their environment altering it as it itself changes.

Soil can be described as a self-organizing system because it dissipates entropy as a result of its development or organization. Pedogenic processes include textural differentiation and the oxidation of organic carbon. Soil-forming processes also include the development of peds (soil aggregates), weathering of clays and sesquioxides from mineral sediments, occlusion and fixation of phosphates and nitrates as well as other organic-derived material, expulsion of wastes such as carbon dioxide, ammonium, and various salts and metals (either as evolved gases or as leachates), and many other processes that together define and create the individual soil.

The soil's exchange of information, energy, and matter with its environment occurs primarily at two spatially distinct locations. At its lower boundary, where it interfaces with parent material, bedrock, and/or groundwater systems, entropy is exported in the form of leachates. In arid regions, greater amounts of water may enter the soil system from the lower boundary than from the upper one [109]. Gases, organic matter, climate, and other biochemical and biomechanical exchanges with the environment occur at its upper boundary where soil interfaces with the atmosphere. Between these two boundaries soil is self-organizing.

A wide range of authors from various scientific disciplines have identified and reported on self-organizing systems. Maturana and Varela [66] use the term *autopoiesis* to define a self-organizing system as a network of production processes where the function of each component is to participate in the production or transformation of other components in the network. The various pedogenic processes presented above describe the production and transformation of components of the soil network.

Autopoietic systems are characterized as dissipative structures that form in a particular way:

1. They are linked with their environment by energy exchange that permits the maintenance of the structure far from equilibrium.
2. They include a large number of chemical reactions and transport phenomena, the regulation of which depends to a high extent on nonlinear factors of molecular origin (e.g., activation, inhibition, direct autocatalysis).
3. They are in high nonequilibrium from the point of view not only of energy but also of matter exchange, since the reaction end products are either eliminated from the system or transported to other locations to fulfill their functions there [110].

Margulis and Sagan [111] describe biotic organisms as autopoietic entities metabolizing continuously, perpetuating themselves through chemical activity and the movement of molecules:

> Autopoiesis entails energy expenditure and the making of messes. Autopoiesis, indeed, is detectable by that incessant life chemistry and energy flow which is metabolism [111; pp. 17–18].

Similarly, the Russian geologist Vladimir Vernadsky, founder of biogeochemistry and a student of Dokuchaev, proposed five functional characteristics that define life processes. These are geologically expressed in a recent translation of Vernadsky's classic *The Biosphere*:

1. A living process is energetic, obtaining and using energy from its environment.
2. A living process is concentrative, accumulating selective elements.
3. A living process is destructive, changing the nature of elements within its domain.
4. A living process is medium-forming, transforming and organizing its local environment.
5. A living process is transportive, moving elements within, as well as into and out of, its domain, without reliance on gravity [112].

In soils, these processes can be interpreted as a kind of metabolism. The processes describe the production and transformation of components of the soil network and the elimination or translocation of waste products to other locations. For example, textural differentiation, available manganese, and the oxidation of organic carbon change together in space and time. The argillic (clay) and stone line horizons, whether formed at the B/C interface or at the surface, demarcate a domain of metabolization where manganese is bio-sequestered for use in degrading the organic carbon.

Development of soil horizons by biochemical and biomechanical means effectively sets limits on the domain of pedogenic processes. These concurrent processes reflect an actively self-organizing system [113]. Pedogenesis is the result of a complex network of production processes whose visible or measurable results include organic carbon diagenesis, the development of soil peds, organics complexed with clays and sesquioxides, and the weathering of clays and sesquioxides. Additional soil metabolic processes include the occlusion and fixation of phosphates and nitrates, elluviation and illuviation, and various other processes that together define and create the soil body [114].

Metabolic processes are not unidirectional but consist of both disassembling (rotting and weathering) and reassembling (bonding together) materials. Organic carbon is both degraded (mineralized into carbon dioxide) and transformed into other organic molecules that are preserved through various biotic and abiotic pathways. These transformations occur as a fundamental process of the soil [115–125] and would not likely take place out of this context. A strong connection exists between the biotic and abiotic soil subsystems such that the existence of one can be conceived as developing a dependence on the processes of the other. Together they form a part of the network of production processes that constitute the organized entity as a whole.

The constituent parts of soil can be inventoried separately, as in traditional pedology and biology, in an anatomical format, but explanations about their synergistic functioning require a more holistic perspective. Such explanations of dynamic systems require a physiological format, as first suggested in a 1789 speech by James Hutton, the father of modern earth sciences [125]. Retired US President James Madison also proposed a system of agricultural reform in 1818 consistent with Hutton's dynamic physiological perspective [126]. From the beginnings of earth science to the present, the scientific literature is filled with the call for a physiological perspective [112,127–134].

Explanatory Soil Science

Explanatory soil science begins by describing the soil system as a network of production processes in which the function of each component is its participation in the production or transformation of other components in the network. If self-organizing systems are conceptualized as constitutive processes (actions) rather than the materials (things) from which they are formed, then soils may be thought of and understood through a common set of processes that are uniquely structured and expressed based on the individual soil's history of perturbations. The processes of soil development consist of a complex network of production processes that together increase the organization of the soil system by dissipating entropy back into the environment.

A soil's ongoing organization is driven by its incorporation of first-order perturbations and its adaptation to second-order perturbations. This energetic relationship between a system and its environment filled with perturbations provides sufficient explanation for why systems are self-organizing, self-maintaining, and self-replicating. There is no need to presuppose a vital force (*vis vita*) or a prime mover [47].

As a complex system, soils exhibit self-organization and self-definition. The soil body is not so much created by the five factors of soil formation as traditionally presented in soil science. Instead, the soil body creates itself through its interaction as a self-organizing and self-maintaining system with its environment that includes these factors.

It is not that perturbations cause soil changes, but rather that perturbations allow emergent properties as new patterns develop in the relationships between components of the system, and with the system's environment. In the biological world, many characteristics that superficially appear as adaptations in fact originate in *exaptations*, evolutionary changes that themselves were initially incidental to the adaptive processes at work but that subsequently proved adaptive in other ways. Another way of putting this involves the functionalist confusion of consequence with cause: birds don't fly because they have wings and feathers; they have feathers for thermoregulation and they fly to escape predators or to catch prey. Living biotic systems are self-organizing systems, and as such the workings of biotic systems are normally described in terms of processes. I do not argue that soils are living biotic organisms. Rather, it is my intent to demonstrate that self-organizing systems may also describe abiotic living systems. While it is traditional to perceive abiotic systems as external to living systems, this has been based on a reductionistic definition of life [135]. If instead the processes that are manifested by a system, biotic or abiotic, are considered, then a commonality is found linking both dynamic systems.

The recurrent perturbations issuing from the environment trigger the complex relationships between the components of the soil system. For example, while rainfall supplies much of the fluid required by a soil to perform its metabolic functions, the soil and not the environment defines how that rainwater is used, in what order, and the rate of these functions. Soil texture and mineralogy, reactivity, and biotic makeup place constrains on these functions, and thus each soil makes use of rainwater uniquely due to its specific history of perturbations. This self-definition is evident when a change in climate (second-order perturbation) triggers a structural change redefining how these metabolic processes are actualized. A fundamental interdependence between structure and function exists for self-organizing systems. "The spontaneously emerging structure corresponds to the systemic function" [110; p. 41].

The structure of a self-organizing system is determined by the actual components, including their properties, but more importantly, by the relationships that exist between the components. Self-organizing systems are defined by their organization, not simply by a listing of their component properties. Rather, self-organizing systems must be explained in terms of the relationships that occur among their components. Rocks and rock sediments, organic matter, mineral salts, and carbonates, as well as macro- and microorganisms, compose in part the specific components of the soil, but the relationships that exist between these components will be based on each soil's unique history of perturbations.

Summary

Modern soil science developed primarily to address societal needs for food production. As such, modern soil science is more an adjunct to plant science than a focused study of soil. The dominant paradigms and models employed by modern soil science are based on an epistemology that seeks to reduce a system to an inventory of its constituent components, failing to adequately treat constitutive processes such as perturbations. Understanding soils in terms of their perturbations require a process approach focused on the interactive behavior of soil's constituent parts as they relate to the soil's environment.

Perturbations are ubiquitous at all spatial scales and occur and have occurred throughout a soil's history. Perturbations come in all sizes and occur at varying frequencies, and their collective effect on a soil constitutes the soil's history. First-order perturbations are common and gentle, and a soil is organized to depend on these first-order perturbations to provide dynamic potential and kinetic energy to the soil system. Second-order perturbations are more severe but less frequent. Second-order perturbations challenge a soil's existence as an entity, requiring that it adjust its structure—the patterns of dynamic relationship between its constituent parts—to maintain its organization and its identification as an entity. New emergent properties result from these changes in a soil's structure as a result of second-order perturbations. Third-order perturbations are catastrophic to a soil, destroying its organization. While a new soil might develop from the constituent parts of a soil destroyed by a third-order perturbation, it would have a different organization and thus would be a different entity.

Soils viewed in terms of their perturbation history are abiotic self-organizing living systems that maintain themselves through metabolic processes comparable to biotic living systems. As such, soils can be seen from the perspective of their physiology as well as their anatomic makeup. In the following chapter I explore soil physiology, distinguishing between those pathways that are biotic and those that are abiotic in nature.

4 Soil Physiology: Processes of a Dynamic Process-Oriented Soil Science

In the previous chapter I introduced Vernadsky's five functional biogeophysical characteristics of life processes. In this chapter I translate these to five essential physiological soil processes that themselves are tied to the 13 paired processes used in traditional soil science presented in Chapter 2, thereby beginning the process of integrating a process-oriented soil science with traditional soil science. The five essential soil processes may be characterized as follows:

1. Metabolic processes include the creation of new minerals and compounds (anabolic) and the weathering out of older minerals and compounds (catabolic): bringing things together and taking things apart.
2. Respiratory processes include the exchange of gasses between the soil and its environment, principally the intake of oxygen and the exhalation of carbon dioxide, ammonium, and other waste gasses.
3. Reproduction in the broad sense is the replication and creation of new soil components.
4. Internal and external information communication systems define how soil is informed about its environment and how it responds to that information.
5. Biodynamic feedback systems provide chelates, enzymes, and other catalysts for physical and structural change in various soil components.

I discuss these soil physiological processes in turn below.

Metabolic Processes: The Digestive System of Soil

Common to self-organizing systems are processes characterized by the assembly of more complex structures from simpler ones (anabolic), and processes characterized by the dissembling of complex structures into simpler ones (catabolic). These two metabolic processes are found in both biotic and abiotic systems. Soils then self-organize, self-maintain, and self-replicate by means of anabolic and catabolic processes. Understanding these various processes is assisted by a physiological assessment, not metaphorically but actually, of the relationships between the system's components.

The energy released as soil microbes degrade organic compounds, for example, is used in part by the microbes and in part in the restructuring of the soil. Restructuring occurs through textural sorting; expansion of the domain of metabolization where digestive processes such as weathering of rock minerals and their reformation into clays take place; boundary formation and maintenance; and waste elimination. These restructuring processes further facilitate and increase the ability of the soil microbes to degrade organic and inorganic compounds in a mutually beneficial feedback between interrelated independent systems.

The domain of metabolization, when freed of excess clays, salts, sesquioxides, and coarse particles, maximizes soil porosity and minimizes soil bulk density. Boundary formation and maintenance protect the area of metabolization from erosion at the surface and from waste backup due to groundwater hydraulic pressure at the base. Erosion and basal hydraulic pressure may be considered as potential third-order perturbations, or at least second-order perturbations that may have led to the emergence of the boundary in the first place. Waste elimination increases the rate at which the biodegradation of organic compounds can occur by removing salts and sesquioxides that otherwise would inhibit both the biotic and abiotic subsystems operating in soil digestive processes.

The metabolic processes of soil follow two distinct but related pathways. One involves organic carbon, and the other silica. Both pathways include catabolic (decay and weathering) and anabolic (humus and clay formation) processes.

Biotic living systems metabolize matter from their environment, transforming that matter into energy and essential building blocks for growth, and dissipating entropy for their own structural change and for the maintenance of their organization as individual entities. These same processes are evident in soils. Soil metabolic processes are not unidirectional but include both the disassembling (rotting and weathering) and reassembling (synthesizing and bonding together) of various organic and inorganic materials.

Organic carbon is degraded and transformed into other organic molecules through biotic and abiotic pathways. These transformations occur as fundamental soil processes [115,118–124,136] and would not likely occur out of this context. New organic carbon, in the form of roots and decaying macro- and microorganisms, enters the soil system within the near-surface soil environment [137–140]. This raw organic matter is then transformed into humus through the processes of humification, decomposition, and synthesis [141] through biotic and abiotic pathways.

The relationship between the biotic and abiotic pathways may be conceptualized in terms of endosymbiosis between two different self-organizing systems, similar to the relationship between the termite and the bacteria living in its hindgut. The existence of one depends on the processes of the other to such a complete degree that they together form a part of the network of production processes that constitute the organization of the entity as a whole.

Soil organic carbon becomes more recalcitrant, less labile, as it moves downwards in the soil profile in relationship with textural differentiation. Textural sorting runs concurrently with the translocation of soil organic carbon. Similar correlations with depth can be found for free iron and other sesquioxides [142].

Clay metabolism has its roots as an independent prebiotic metabolic strategy that underwent changes subsequent to the advent of self-organizing biotic systems. The resulting changes in the crystalline structure and diversity of clays are adaptations to changes in the soil's environment to include biotic systems. An example of a similar circumstance involves today's atmosphere, now dominated by oxygen instead of carbon dioxide, and the increased organic carbon resulting from the wastes of biological life. The weathering of silicates, such as feldspar and mica, is greatly enhanced in the presence of carbon dioxide as the calcium recombines to form calcium carbonates [143,144], leaving behind silica, which is available to combine with aluminum or other metals to form clays of various kinds [145].

Prior to the emergence of biological (organic carbon-based) life 3.9 billion years ago, when the atmosphere was dominated by carbon dioxide, methane, and water vapor, the catalyst in clay metabolism would have been limited to inorganic hydrocarbon monomers [146,147].

Inorganic hydrocarbon monomers (such as methane, carbon dioxide, and ammonium) can be transformed into more complex polymers through three simple processes: condensation, polymerization, and oxidation and reduction. All three processes involve either the removal or the inclusion of a water molecule through the association or disassociation of the hydrogen (H^+) and hydroxyl (OH^-) ions [148]. The edges of clay crystals possess a strong ionic charge that is referred to as the edge effect. Isomorphic substitution, commonly consisting of the replacement of a magnesium atom for an aluminum atom, or an aluminum atom for a silica atom within the crystal matrix, is another source of ionic charge in the clay crystal [149]. The ionic charge of clay crystals would have catalyzed the association and disassociation processes of the inorganic hydrocarbon monomers. Clay also acts as a catalyst both in the removal of waste products and in the retention of needed minerals due to its colloidal properties [149].

The architecture exhibited over the full range of clay species characterized by isomorphic substitution would have included chemical symmetry. But it is also likely that chemical asymmetry would have existed for specific polymers, based on the unique clay template formed when protected from hydrolysis [150,151]. Research has shown that the crystallization process of some clay types is facilitated by, and may even require the presence of, specific amino acids [151–154]. The oxidation of sequestered humic material within smectite clays gives off heat, leading to the recrystallization of the smectite to illite [151]. It has also been suggested that clay's crystalline properties may have served as a template for organizing inorganic carbon monomers (methane, carbon dioxide, and ammonium) into the first self-replicating nucleotides and amino acids [117,155–157]. Clay's metabolization processes in this sense may be seen as creating (emergent property) a new component (self-replicating organic polymers) for use elsewhere within its organizational system, a fundamental behavior of a self-organizing system.

Numerous materials that enter the soil system through biogeochemical weathering, and the processes of hydration, dissolution, oxidation, reduction, and hydrolysis, are used and transformed during the metabolic process in soils, as is the case with enzymes and vitamins in biological life [158–160]. Multiple forms of sesquioxides within the zone of metabolization as well as the role of salts as enzymes/chelates are essential to soil metabolic processes. The process of podzolization can be used to illustrate this metabolic function [161].

Podzolization occurs in acidic, medium- to coarse-textured soils that are generally characterized by low seasonal temperatures and abundant rainfall. Production rates of organic matter under these conditions are generally greater than decomposition rates, resulting in a net surplus of humic material. The acidic conditions result in the formation of clay–sesquioxide and clay–organic–sesquioxide complexes and their translocation into lower portions of the soil profile where the sesquioxides, including aluminum, iron, and manganese hydroxides and oxides, are released [118,119,162–168]. Sesquioxides are required in small amounts to chelate silicate clay crystallization and organic carbon decomposition, but in large concentrations sesquioxides, particularly aluminum, inhibit organic carbon decomposition and plant growth in low-pH soils [168,169].

Large-polymer humic materials such as humic acid molecules readily combine with aluminum hydroxides and iron and manganese oxides, becoming water soluble. These water-soluble organic sesquioxide complexes leach downwards out of the organic A horizon, resulting in the characteristic bleached (albic) A horizon of spodisols [170]. Aluminum hydroxides oxidize to aluminum hydroxyl ions or integrate into the crystalline lattice of the clays, releasing the humic acid molecule on contact with the less-acidic and silica-rich aluminum-starved clay material in the upper levels of the B horizon. The humic acid molecules, along with the iron and manganese oxides, are no longer water soluble with the release of the aluminum hydroxyl ions and accumulate in the upper B horizon, combining with silica clays to produce the diagnostic spodic horizon of the spodisol [171,172]. As I discuss subsequently, the increased accumulation of clay and organic carbon is evident in OCR data accumulated from archaeological sites.

Various perturbations currently contributing to the process of haploidization may be viewed physiologically as an emergent property of soil: an adjustment, or corrective, mechanism necessary for the efficient management and metabolization of organic carbon. The contrasting vertisol and spodisol soil orders, both exhibiting high levels of perturbation, can serve as examples. Vertisols are characterized as soils lacking evidence of eluviation and illuviation that form in climates marked by seasonal wet and dry cycles. They have a high, poorly weathered clay content, free carbonates, and generally high to very high pH values. Spodisols are characterized by extremes in eluviation and illuviation and are found in wet and generally cold environments. They form in coarse-textured low-pH sediments and are typically marked by moderate to high levels of available sesquioxides.

Concurrent with the process of expanding their domain of metabolism, soils will expel both fine and coarse particles and excessive or waste minerals such as sesquioxides from the area of metabolism. Vertisols and spodisols represent extreme conditions where the removal of inhibitors to metabolization, such as clays, carbonates, and sesquioxides, cannot be efficiently conducted by the soil within the conditions imposed by its specific environment. Complexes of organic carbon are formed by clays and carbonates in the case of vertisols, and atmospheric oxygen diffusion rates are restricted. Complexes of organic carbon and clay are formed along with sesquioxides in spodisols, and due to the low pH value, cool temperature, and high rainfall, are flushed from the area of metabolism (the A horizon) into the B horizon. One of the primary metabolic functions of the soil in both of these cases, organic carbon management and its digestion and metabolization, is not fully realized. Adaptation to this condition results in the emergence of new strategies.

Argillipedoperturbation, the shrinking and swelling of 2:1 lattice clays characteristic of vertisols, physically churns the material within the domain of metabolization, optimizing oxygen access to all parts over time. Free carbonates are leached, 2:1 lattice clays are weathered, and organic carbon is oxidized as a result of this churning, albeit in slow motion when compared to other soil orders such as mollisols. This adaptive digestion structure compares to our own stomach, which mechanically churns its contents to increase consumed organic matter (food) exposure to digestive acids, enzymes, and bacteria.

Bioperturbation, resulting from tree-falls, is characteristic of most (if not all) spodisols, and this perturbation effectively returns the complexed and illuviated organic carbon back into the domain of metabolization. The reduced and chelated sesquioxides are thereby released from their organic complexes, allowing the organic carbon to mineralize (evolve off as carbon dioxide) or to be resynthesized into other humic molecules. The episodic release of the sesquioxides complexing with fully oxidized organic carbon (vitrinite and lignite), along with weathered secondary clays, may later resolve as the pedogenic lamellae common in the subsoil sediments of spodisols. The haploidal process seen in spodisols' adaptive digestion structure compares to a ruminant's stomach, where the contents are periodically regurgitated for additional mechanical and chemical processing.

Data from a study of soil carbon content versus age for a Holocene chronosequence suggest that each of the major soil taxonomic orders will show unique metabolic characteristics [173]. A study of carbon storage in loess-derived surface soils from central Germany similarly suggests that other soil characteristics participating in the metabolic process may be uniquely ascribed at lower taxonomic levels [174].

Individual landforms as well as currently defined soil taxonomic units can be characterized therefore by how the soil metabolizes organics, a process-oriented classification. The assessment of other soil constituents involved in soil metabolic processes would likely provide additional differentiating evidence.

Respiration System

Soils exchange various gasses with their environment in addition to the intake of organic material. Oxygen taken in from the atmosphere is used in metabolizing soil organic carbon [175,176]. Oxygen is also essential in the formation and speciation of clays [151,153,177,178] and sesquioxides [164,179,180], both of which are mediated by biotic and abiotic processes [181]. The intensity (rate of reproduction) of life can be judged by the rate of gaseous exchange, or respiration [112]. Respiratory rates for plants and animals are presented in terms of mols, or millimeters, of carbon dioxide per time, measured in seconds [182]. Respiratory rates for soil are presented in terms of grams of carbon dioxide per time, measured in hours or days [175]. Currently, soil respiration studies measure carbon dioxide as an indicator of overall biological activity in soil, but this may also be refined to characterize abiotic activities.

Free atmospheric oxygen, however, is a relatively recent gas that is a result of pollution by photosynthesizing biotic life forms. The Earth's atmosphere prior to roughly 2 billion years ago was dominated by carbon dioxide [146]. The lack of free oxygen means that iron and other metals in the soil would have existed primarily in a reduced water-soluble form and thus would have been subject to loss from the soil body through leaching. (The potential importance of sesquioxides as catalysts in soil metabolization was discussed earlier with reference to metabolic processes.) Iron and aluminum, along with calcium, also bind phosphorous. Both iron and phosphorous are essential nutrients critical for biotic life in the soil [172].

Sulfur, conversely, is water soluble in its oxidized form. With the advent of an atmosphere containing free oxygen, sulfur would have been quickly lost from the soil through leaching. In humid climates sulfur is present in soil primarily in combination with biologically generated compounds and becomes available to other plants only during the biodegradation of organic matter. The strategy of phytocycling sulfur by plants makes sense from the plant's point of view, but it also proves beneficial to the soil itself. Soil pH decreases as elemental sulfur oxidizes, rendering other elements unavailable for the metabolization processes. Decreasing pH values also increases the mobility of sesquioxides [183].

Reproductive System

Abiotic reproduction will not likely resemble present-day biotic strategies of sexual and asexual reproduction [184]. The RNA, DNA, and protein synthesis normal to biotic reproduction would not exist within the abiotic system. A similar but less "high-tech" system might be expected instead:

> "We should doubt, then, whether amino acids that are so good for making catalysts (given the technology) would have been good for catalysts to begin with. We should doubt whether amino acids or any other of the now critical biochemicals would have been at all useful right at the start" [155; p. 91].

Research into the antiquity of RNA-based evolution provides strong support for the idea that biotic life as known today is a complex system that had precursory less-complex systems. The "RNA world" theory suggests a period of time before DNA-based life where RNA molecules rather than proteins catalyzed and synthesized the necessary biological molecules for reproduction [185,186]. RNA is a variably sequenced polymer that is amenable to the use of templates for self-replication, such as those offered by clay crystals [185]. Arguments against this hypothesis have focused on RNA's lack of chemical diversity, which is believed necessary to sustain life [144].

Studies have also shown, however, that even simplified RNA molecules (consisting of three rather than four ribozyme molecules) are sufficient for these types of catalytic actions [187]. Efforts to define possible predecessors to the "RNA world" have resulted in theories focused on the early polymers of threose nucleic acid (TNA) and peptide nucleic acid (PNA). Both of these are capable of forming stable Watson–Crick pairs. Two of the four key bases of DNA, guanine and cytosine, form in clay-rich solutions [188]. This suggests that organic catalysts even simpler than RNA would have been capable of self-replication. It is possible that RNA may have acquired many of its organizational attributes from these simpler forms [185], and that they may have in turn been informed by the organization and replication processes of clay crystals.

Clay crystals have been suggested as a more likely start of the reproductive process for biotic life [116,117,151,155,177,189–193]. Rates of clay formation, or reproduction, operate at different timescales than biotic reproduction, requiring thousands of years [194] rather than minutes to decades for biotic reproduction [112]. Clay crystals are aperiodic. Errors can be introduced as clay crystals replicate by way of second-order environmental perturbations, resulting in the isomorphic substitution of alternative atoms, often manganese or iron, for atoms of silica. These errors are inherited by subsequent replicates, constituting a mineralogical correlate to a genetic mutation [119,149].

All biotic living systems are genetically linked, as has been shown by the detailed work of Dr. Carl Woese [54]. Thus they share a genetic history that is not generally thought to be shared by abiotic systems like soil. These two systems, however, do share the same broadly defined processes, differing more in scale than in kind. This suggests that what is identified as biotic life may be an emergent property of greater scalar complexity. It is not necessary that soil, as an abiotic system, be genetically linked in the strict sense with biotic life. The characteristic that both biotic and abiotic self-organizing systems share is the property of self-replication. Thus Woese's tree of life may be perceived as embracing all biotic living systems, but it can also be considered as firmly planted in, and as having arisen from, the soil. Self-organizing systems, both biotic and abiotic, may be seen as having a history of increasing complexity in terms of their organization to undertake catalytic processes. The distinction between biotic and abiotic systems becomes one of degree, and not one of kind.

Internal and External Communication System

Mycorrhizal fungi constitute an essential component of soil systems, mediating between plants and soil to maintain the living upper boundary of soil, the vegetation. Mycorrhizal fungi provide inter- and intra-component communication, beyond just individual soil pedons, organizing the distribution of soil and plant resources (organic as well as inorganic) across landscapes due to their extensive aerial coverage.

In Washington state, for example, one individual of *Armillaria ostoya*, a parasite of conifers, is reported to inhabit 6.25 km^2 [195]. A single *Armillaria bulbosa* found in Michigan covers 15 hectares [196]. In a recent Technology, Education, and Design (TED) lecture Stamets [197] mentions a single 2,000-year-old fungal species covering 890 hectares.

Mycorrhizal fungi redistribute resources throughout their large mycelial network of hyphae, mobilizing carbohydrates and essential nutrient elements such as nitrogen and phosphorus, as well as various earth metals from a wide range of spatially separated substrates of different qualities, and translocate these obtained resources throughout their mycelium, in turn making these nutrients available to plants and soil processes as well.

A significant proportion of calcium in some tree species (particularly conifers, beech, and birches) growing in calcium-poor acidic soil has been shown to come from apatite, a soil mineral mined by fungi that live on tree roots [198]. Fungi have similarly been shown to influence phosphorus solubilization and pH changes in the soil [199]. Large quantities of heavy-metal contaminants, including zinc, copper, mercury, lead, and cadmium, as well as radioactive pollutants in soil, are also metabolized by various species of mycorrhizal fungi [196,200–203]. Fungi, further, are capable of incorporating complex polycyclic aromatic hydrocarbons (PAHs) from contaminated soil [204].

Research on fungi has been predominately focused on the various plant–fungi interactions [205]. A symbiotic relationship exists between most plants and mycorrhizal fungi, with mycorrhizae receiving plant-produced sugars and (some speculate) essential growth regulators, in exchange for phosphorus, potassium, copper, and zinc, nutrients that are not readily soluble and available to plant roots [154,206].

The rate of information and matter exchanged through this *neural*-net of the soil has not, to my knowledge, ever been directly measured. An analog model, however, may be employed using the visible above-ground fruiting bodies (mushrooms) of specific species as they spread across a forested area. The bay bolete (*Boletus badius*), which is mycorrhizal with white pine and hemlock [207], is common in the temperate mixed pine hardwood forest near my home and presents an easy and convenient case study. This mushroom is easily recognized and not likely confused with other fungi in bloom at the time. The only mushroom that is the same size, has the same color and shape, and is out at the same time of year in this area is a gilled mushroom. The bay bolete is a polypore sponge mushroom. Other polypors, such as the bitter bolete (*Tylopilus felleus*), resembling the bay bolete, become common during the course of the season, but they exhibit easily recognizable distinguishing features such as pore and spore color, and size.

The bay bolete fruiting body emerges in late June and early July when the weather warms to summer conditions that commonly trigger late-afternoon or early-evening thunderstorms. The first fruiting bodies usually emerge at a sharp ecotone with a different forest community or a sharp change in slope due to incised drainages. Each individual mushroom generally lasts only about a day, due to predation by slugs, squirrels, and epicureans. New fruiting bodies emerge with each successive day, appearing within a few meters of the first bloom and expanding out across the full reach of the forest patch occupied by this mycorrhizae system.

A reciprocal wave of mushrooms returns back across the area to the point of its initiation after reaching the outer edge of its forest patch, a sexual spasm rippling through geoderma. The reciprocal wave, just as a ripple in a pond, is lower in amplitude (fewer mushrooms) and takes a little longer to traverse the terrain. The bay boletes continue to emerge throughout the summer months and well into the fall, but they appear randomly within the forest patch and are fewer in number. In my harvest (study) area that measures just under a half-acre, the rate of spread in the initial wave is roughly 200 m over 8 days, with the return wave taking only 9 days. This is a blinding speed for a system operating at the pedogenic scale.

Clearly, the soil communication system, its neural-net, is a biotic system, fungi. To date, I have not been able to discern any parallel abiotic system for internal and external communications for soils, although redox potentials for sesquioxides and clays may provide such. Future research on soils as dynamic process-oriented systems may eventually uncover novel abiotic systems not imagined at present. Regardless, that fungus emerged as a biotic system exhibiting communication behavior interceding between plants and soil is significant in its own right. Much research has been undertaken documenting how these behaviors benefit plants; little has been done to understand and explain the adaptive values for soil. The fungi's intercessions promoting stable diverse plant communities that operate as a membrane between the atmosphere and the soil's metabolic core, and mycorrhizae's capacity to improve and maintain soil structure are clear examples of soil physiology.

Biodynamic Feedback Systems

The endocrine system is critical to living biotic self-organizing systems. In higher vertebrates it is less an independent subsystem of the body than a biodynamic feedback mechanism attached to the neural and vascular system. Stimulated by chemical signatures, primarily hormones in the circulatory system and neural endorphins from the brain, the hypothalamus, pituitary, and adrenal glands secrete hormones to activate, stimulate, or repress the actions of various organs throughout the body [208].

Similar processes are evident in self-organizing soils. The free oxygen in the atmosphere and in the soil "was put there by manganese" [209]. Manganese is also involved in the degradation of organic carbon in soil and in the reduction of nitrate. Manganese oxides and other free radicals act as catalysts in that they are reformed and recycled. Figure 4.1 shows how manganese values and their relationship to percent total carbon help define active zones of metabolization.

The fruiting bodies of mycorrhizae mushrooms, as well as cryptogamic crusts and lichen, produce a wide array of complex polymers, including various proteins, many of which are classified as hormones and endorphins, as well as alkaloids that may function in a similar fashion on various soil processes. Mycorrhizae hyphae, as well as cryptogamic crusts and lichen, also effect the microaggregation (peds) of soil particles by secreting the protein glomalin along with other organic and inorganic chemicals that bind soil particles. Various *Boletus* species produce complex polymers, such as ephedrine, an aromatic amine, and muscarine, a cholimimetic alkaloid that may improve the weathering of silica from quartz and the production of silica and hydrous oxide clays by affecting the soil pH. The rate of enzyme-catalyzed hydrolysis in animals is fastest at about pH 7.4 [210], but this optimal value may be specific to the hydrolysis of organic compounds. Chemical weathering of silicate minerals is primarily the result of the hydrolysis of silicic acid and the removal of associated bases, both requiring pH values somewhat below 7.

Soil Depth	pH	%Organic Carbon (LOI)	OCR Date	Very Coarse	Coarse	Medium	Fine	Very Fine	Coarse Silt	Fine Silt	Sample Id	%Oxidizable Carbon (WB)	OCR Ratio	Mn
5.5	4.0	4.924	382	0.705	0.542	0.527	8.077	50.942	13.983	25.224	5647	1.91	2.58	3.91
10.5	3.9	4.652	353	0.439	0.551	0.565	8.077	58.400	12.432	19.536	5648	1.95	2.39	2.18
15.5	4.0	4.467	348	1.591	0.601	0.520	8.409	52.170	14.498	22.211	5649	1.92	2.33	2.18
20.5	4.1	4.433	354	0.536	0.528	0.489	9.047	50.192	15.470	23.737	5650	1.92	2.31	2.73
25.5	4.1	5.140	546	0.732	0.657	0.600	8.744	53.390	13.097	22.780	5651	2.08	2.47	1.76
26.5	4.2	4.957	601	1.500	0.841	0.834	9.245	59.236	11.384	16.959	5652	1.79	2.77	0.13
30.5	4.1	4.513	796	1.587	1.105	0.855	10.213	61.036	10.276	14.928	5653	1.43	3.16	0.07
35.5	4.2	3.691	1108	1.737	1.275	0.916	10.371	60.737	11.499	13.466	5654	1.10	3.36	0.01

Figure 4.1 How manganese values and their relationship to percent total carbon help define active zones of metabolization.

Mycorrhizal fungi also act as a sort of gatekeeper, allowing beneficial soil organisms to pass while preventing harmful pathogens from crossing over to plants [196,211,212], functioning as a kind of autoimmune system. Mycorrhizae actively intercede in the development and composition of plant communities by inhibiting the growth of annual weeds that do not provide efficient protection against erosion while at the same time facilitating the growth of perennial grasses and shrubs that are significantly more effective at protecting the soil [213,214]. In aridisols, cryptogamic crusts containing a fungal component excrete polysaccharides that help bind soil particles together, and may also influence (chemically and/or physically) the germination of exotic annual (invasive) grasses [215,216]. Fungi also control micro- and macro-invertebrates in the soil through the production of natural insecticides. The birth of the germ theory in biology came about with the identification of *Beauveria bassiana* as the cause of "muscardine disease," which was ravaging the silkworm industry of Europe in the early 1800s [214]. Ongoing research is isolating and identifying a wide range of new antibacterial, antifungal, and anticancer compounds in numerous mycorrhizae fungi [217–221].

Mycorrhizal fungi interact with a wide range of other soil organisms in the root, in the rhizosphere, and in the bulk soil, stimulating or inhibiting their functions in competitive and mutual relationships [222,223]. One study has shown that fire ants change the physical and chemical components of soil in order to influence the soil-fungi composition in their mounds. The differences between fungal communities in soil from native and nonnative ant colonies might indirectly influence ant-mediated seed dispersal by affecting seedling survival [224].

Summary

Soils exhibit dynamic behaviors comparable to self-organizing biotic entities when seen from a dynamic systems perspective. Soils metabolize carbon and silica as well as other elements weathered from plant and mineral decay. Soils create boundaries defining their zone of metabolization. Soils process and eliminate waste. Clay production and speciation, as well as numerous other metabolic processes, are determined by the soils' organization. Soils respire, exchanging gases with their environment to facilitate various metabolic processes such as weathering, and clay formation and speciation. Soils reproduce through the growth of clay crystals, which, being aperiodic in structure, are capable of introducing replicatable error. Soils, consisting of parent material and biotic components, communicate with their environment and, more importantly, between its own various components. Lastly, soils display feedback systems whereby their organization is maintained through internal stimulus–response behaviors.

With some of these physiological processes two distinct pathways, one biotic and the other abiotic, are evident, suggesting an evolved relationship between the two dynamic systems. This combination of biotic and abiotic metabolic pathways can be conceptualized as resulting from a process similar to, and perhaps antecedent to, that of endosymbiosis in biology. Endosymbiosis results in the combination of two distinct species, usually one predator and the other prey, combining to form a new species, with the original species becoming interacting components whose effects are mutually beneficial rather than antagonistic. This theory was proposed to explain the evolution of nucleated single-cell bacteria from unnucleated protist [205]. If so, it becomes possible to model early biotic systems as an emergent property of a prebiotic soil, with the properties of these self-replicating organic molecules enhancing various soil metabolic processes.

Some modern soil processes do not appear to have a clear abiotic counterpart (specifically, communication and biofeedback systems). These systems likely emerged from the biotic component of the soil system and were maintained as beneficial to other soil physiological processes. However, with further physiological research into soils as self-organizing systems, prebiotic correlates may yet be discovered.

That soils exhibit processes comparable to biotic organisms does not mean, however, that soils and biotic organisms are the same at all levels of abstraction. The dynamic processes exhibiting self-organization, self-maintenance, and self-replication may be compared in kind, but timescales and degrees of complexity between the two systems are vastly different. What is important is the fact that these processes are common to all self-organizing systems, both biotic and abiotic. Soils, both at the pedon and landscape scales, can be assessed through a physiological perspective that describes behaviors and processes, as well as through an anatomical perspective that describes attributes, components, and characteristics.

5 The OCR Procedure as Applied Dynamic Process-Oriented Soil Science

The Initial Iterations of the OCR Formula

The previous two chapters introduced some novel explanations for soils and how they behave. If left as is these explanations would be merely speculations having little use or importance to soil science. The methods and procedures of physiological assessment of a dynamic system are different from those of an anatomical description of the components of that system. It follows, then, that new and different approaches to soil analysis will be required as part of incorporating an explanatory component into modern soil science.

Many of the ideas and theories expressed in the previous two chapters are the product of my attempts to understand and explain the apparent anomalous results obtained from the Walkley-Black and Ball Loss on Ignition procedures when applied to subsurface and archaeological soils. The ensuing development of the OCR procedure, which combines these two organic carbon analyses to calculate age estimates for soil organic carbon, has also proven to be a valuable tool for the physiological assessment of soils as dynamic systems. As such, it is important here to detail the development of the OCR procedure and its application as a diagnostic tool for soil physiology through various case studies.

The OCR Procedure

As noted at the outset, my interest in soils was sparked by an initial analysis of three archaeological hearths using the Walkley-Black and the Ball Loss on Ignition procedures, both of which measure soil organic carbon. I expected based on the available literature that the results would be similar, with the Walkley-Black procedure recording 75% to 80% of that recorded by the Ball Loss on Ignition procedure. Instead they varied by as much as one order of magnitude. Shortly after discovering these seemingly anomalous results, radiocarbon ages obtained on charcoal samples from these features arrived. The noted deviation in my results appeared to correlate with the ^{14}C age of the samples. The older ^{14}C dates correlated with larger discrepancies between the two organic carbon procedures. A linear correlation between only three points is not significant, so I obtained additional samples from other archaeological sites. My goal was to determine the factors that affect the correspondence between the ratios of total organic carbon to readily oxidizable carbon and their relationship to the age of the sample. The different values obtained by the Walkley-Black procedure compared to the Ball Loss on Ignition procedure are due to the inability of the wet combustion procedure to oxidize the more biologically inert forms of organic carbon, which were found to increase over time.

But the apparent linearity diminished with each additional sample, indicating that other factors beside time were likely influencing the results. Additional samples were obtained from soils with different textures and from different depths. Soil texture and depth affect the diffusion rate of moisture and atmospheric gases [225,226] and thus the rate of biochemical reactions. The linear relationship between the variation in organic carbon results and age returned when these variables were factored into the equation.

I obtained more samples over the ensuing years from different soils, in different regions of the country, and from sites of different ages. I found that temperature, rainfall, and soil reactivity (pH) needed to be factored into the equation. Temperature and rainfall affect soil development by reducing soil pH, increasing the rate and depth of leaching, and increasing rates of organic matter decomposition [30]. The speed of chemical reactions increases by a factor of two to three for every 10 °C rise in temperature [227]. Concurrently, organic carbon turnover rates decrease as soil pH values drop due to two separate processes. Microbial activity decreases as soil pH drops and, with pH values below 6.0, soluble and exchangeable aluminum increases and forms aluminum–organic carbon complexes through the various processes of solubilization, condensation, precipitation, aggregation, and occlusion [228]. I discovered that the total percent organic carbon also needed to be considered as a variable because soil microbial makeup and populations depend on the concentration of total percent organic carbon as a potential food source.

The inclusion of these variables effectively reflects the traditional five-factor model of soil formation: climate, organisms, relief, parent material, and time [28,30,229]. As I have emphasized above, the five-factor model is the dominant and historically the only conceptual model governing soil science; it is thought to sufficiently explain all soil formation and soil physical geographical phenomena. Pedagogically, it is the first concept (or one of the first) introduced to the student of pedology and physical geography, and it informs all research concerned with soil genesis and classification.

My initial evaluation of these early samples collected from around New England showed a high degree of variability. When expressed as a ratio of the two procedures (Loss on Ignition/Wet Combustion), the variability ranged from 2.53 to 7.12, with a mean of 3.91 and a standard deviation of 1.09 [230]. I began using this ratio as an index for characterizing soil organic carbon and its age, and named this analytical approach the Oxidizable Carbon Ratio (OCR) procedure. The implications of the OCR variability suggest that the charcoal found in soils is not immune to biochemical alteration over time, as was generally then believed [231], but rather charcoal exhibits a high degree of variability due to biogeochemical degradation [230]. A recent examination of modern and fossil charcoal has demonstrated structural and diagenic change over time [232].

I ran Pearson correlation statistics using these initial data between the OCR and each of the variables and found relatively low positive or negative correlations (Figure 5.1). Multiple linear regression calculations comparing the OCR to the

	Time	Depth	% Total Carbon	Climate	Texture	pH
OCR: r =	+.35	+.31	+.21	−.37	+.24	+.06

Figure 5.1 Correlation statistics between the ratio of organic carbon results and other factors.

combined variables, however, yielded a moderately strong correlation (Pearson's $r = +0.63$), suggesting that an interdependent dynamic in the system was being expressed. Statistical significance tests were not performed at that time. The variability in the OCR ratio was determined to be the result of all the factors in combination, and not just time alone.

All samples collected and analyzed for this initial study were of known age. Older samples were associated with ^{14}C dates, and the more recent samples (primarily from 19th- and 20th-century house fires) were dated through newspaper or reliable eyewitness accounts. Depth, percent total organic carbon, and pH were measured values. Climate and texture were ascribed ranked values in the early studies [230]. These variables were later changed to measured mean temperature in degrees Fahrenheit, measured mean rainfall in centimeters, and mean weight of the combined textural classes [233]. The Fahrenheit scale, although not standard for most scientific measurements of temperature, was employed to accommodate the more northern regions, where mean temperature expressed in degrees centigrade would require employing negative numbers, needlessly complicating computations and creating confusion for archaeologists providing and using these data [233].

Specific analyses for the other variables included results for total organic carbon, determined by the Ball Loss on Ignition procedure [2], readily oxidizable carbon determined by the Walkley-Black Wet Combustion procedure [3,4], measured depth below surface, texture determined by USDA standard mesh screen sizes, mean annual rainfall and temperature based on National Oceanic and Atmospheric Administration (NOAA) narrative summaries for the period 1941 to 1975 [234] or other suitable long-term average records, and pH as determined from a 1:1 paste of soil and water. I elected to use the raw, unconverted, organic carbon results from the two carbon procedures in the initial experiment as the contents of these soils were dominated by charcoal, which contains a significantly greater percent carbon than does raw organic matter.

I formulated an equation at this point that would relate the influence of these interdependent variables to the OCR. Various approximations comparing addition and subtraction versus multiplication and division were developed and tested for highest r values and resulting population distribution statistics of kurtosis and skewness. I found that the relationship between the variables was not cumulative but multiplicative; the dynamics of this system are a result of positive and negative feedback between the variables.

The formula that best explained the data was:

$$\frac{OCR \times Climate \times Depth}{Texture \times \sqrt{\%Carbon} \times \sqrt{pH} \times Time} \times f$$

where f represents a residual factor not accounted for by the other variables. The formula was later recalculated to include the change of climate and texture to mean temperature, mean rainfall, and mean texture. The factor f was then solved from the sample data and found to equal 14.4888. Next the equation was solved for the variable time, and expressed as OCR_{DATE}. This final formulation showed a high correlation ($R^2 = 0.95$) with preexisting dates obtained independently.

$$\text{Time} = \frac{\text{Mean Texture} \times \sqrt{\%\text{Carbon}} \times \sqrt{\text{pH}}}{\text{OCR} \times \text{Depth} \times \text{Mean Temperature} \times \text{Mean Rainfall}} \times 14.4888$$

I then obtained 20 additional samples from undated archaeological hearth features that contained temporally diagnostic artifacts. I used these additional samples to independently test the formula. The resulting data were entered into the deduced OCR_{DATE} formula to generate age estimates. All 20 fell within the expected temporal range of the artifacts [230].

The OCR_{DATE} formula was developed to explain unexpected analytical results on the nature of soil organic carbon. The OCR_{DATE} formula, however, was also discovered to be a potentially useful estimate for the temporal factor in soil development. Jenny's factorial model has been valuable for evaluating the effects of climate, organisms, relief, and parent material, but the effect of time on soil formation has been difficult to determine. Dendrochronology [235–237] and ^{14}C dating [238–240] have been used for calendrical and numerical dating of soil processes, but the former requires the presence of trees with determinable tree ring chronologies and the latter is costly due to fractionation and sample preparation requirements. Additionally, ^{14}C dating assumes a closed system free of outside contaminating influences. Soils are most decidedly open systems. Open system "errors" in radiocarbon dating are controlled by thoroughly "cleaning" samples through the use of vigorous pretreatments to remove "contaminating" younger and older carbon [241].

Several other adjustments and considerations are required for radiocarbon dating due to the fact that earth systems are open systems and do not necessarily follow linear expectations. Short- and long-term variations in atmospheric ^{14}C concentrations exist throughout time due to changes in the amount of cosmic radiation that enters the earth system from space (the de Vries effect [242]). The distribution and storage of carbon is not uniform throughout the earth system (reservoir effect). Furthermore, these reservoirs are themselves affected by their environmental context. Fractionation, the propensity of different plant species to incorporate differing ratios of the carbon isotopes, coupled with soil microbes that differentially remove either the heavier ^{14}C or the lighter ^{12}C isotope from samples, exemplify this open-system variability. Anthropogenic post-16th-century burning of fossil fuels has disproportionately added old carbon (radioactively inert) to the atmosphere (the Suess effect [243]), and nuclear testing, which has increased the amount of young carbon (^{14}C) in the atmosphere, and the atomic bomb effect [244] are further open-system problems that must be taken into consideration when conducting ^{14}C analyses. Further discrepancies exist between the time that atmospheric ^{14}C enters the organic sample, and the event, such as a hearth being dated (the pre-sample-growth error [245]).

Studies on the temporal factor in Jenny's model without numerical dating procedures have been primarily limited to coarse-grained relative dating [69,194,246,247]. The OCR_{DATE} equation is a formulation of data obtained from inexpensive soil analytical procedures commonly run in the course of soil characterization studies.

Of equal importance, the OCR formula represents the first, and still only, actualization of Jenny's [30] theoretical model [248]. The OCR formula includes climate with mean temperature and mean moisture values; organisms with percent total organic carbon, depth, and pH values; parent material with soil texture and pH values; and time with the calculated OCR_{DATE}. The development of the OCR formula occurred using topographically level areas, and relief is not included [249]. The inclusion of relief into the OCR formula, however, would only require obtaining an appropriate selection of samples from various slopes, with the other four variables controlled by their relationship in the current OCR formula.

Evolving Applications of the OCR Formula

Initial analyses suggested that the reason why the Walkley-Black and Ball Loss on Ignition procedures yielded variable results for the organic carbon was due to time [230,232,249–252]. These initial studies using the OCR procedure relied upon the traditional five-factor model of soil formation: climate, organisms, relief, parent material, and time [28,30,229].

Reports on the variable results obtained for the organic carbon by these two procedures since 2002 note spatial changes in textural differentiation and available manganese distribution concurrent with the changing values for organic carbon at the local site-specific pedon scale. Furthermore, these coupled changes are related by mild perturbations and demarcated by greater perturbations that are ubiquitous at the global scale [253–256]. The five-factor model, where soil is simply a product, cannot explain these coupled changes, affecting and affected by each other, within a dynamic context. Their explanation rests in understanding soil as a complex system affected by and responding to an environment filled with perturbations.

The OCR formula was initially constructed from a small dataset of 48 samples. Data obtained from subsequent samples have not required adjustments to the order or organization of the initial equation. The calculated constant has become increasingly more precise and accurate, as measured in comparison with independent numerical chronometric ages. Over 7,000 samples have been accumulated over the past 25 years, and the calculated accuracy and precision of OCR formula continue to improve, suggesting that the order of variables in the equation is inconsequential as long as relationships are expressed in terms of an augmentative numerator and denominator. The architecture of the OCR formula allows positive, negative, or buffering feedback behavior between attributes to be expressed rather than masked. This observation would suggest that the OCR formula could be used as a basic model for the physiological assessments of other soil constituents, for example, free iron and specific pedogenic clays that also move through the soil over time.

6 Applications of OCR Dating

I argued in the previous chapters that soil can be viewed as a complex evolving self-organizing system, open relative to energy and matter. Soils are not static but are constantly undergoing biochemical and physical changes according to a specific organizational design. Their function of managing and metabolizing organic carbon as well as clays and their consequent throwing off of carbon dioxide, heat, and other wastes conform to physical laws for dissipative systems [105]. The dynamic soil system as presented in the previous chapters is a maze of interactive dependent and independent components that cannot be fully understood through a descriptive inventory of constituent parts.

A soil's history is composed of numerous punctuated disruptions that may be defined as moderate or second-order perturbations. Moderate perturbations affect the structure (pattern of relationships between constituent attributes) of the soil system but not its organization. Here, I examine several complex soils, dissecting them according to the temporally independent perturbational events that define each soil package as a unique entity. These structural changes are often invisible as the soil body adapts to the perturbations, conserving its unique organization. The effects of these perturbations, however, do remain evident in the soil as detectable structural changes and are demarcated as event horizons or soil packages. The OCR procedure measures physical and chemical changes that have been actualized and that continue to occur in soil. Time is inferred from these changes.

Several case studies follow. These detail the use of OCR analysis and specify the dynamic process-oriented interpretations drawn from the results. All of these studies were conducted during archaeological research and were undertaken to establish various contextual concerns of actual sites. These case studies are here loosely grouped as geomorphic, environmental, and cultural studies.

Geomorphic Studies

Paso Otero 5, Argentina

River systems and their distribution of sediments are extremely sensitive to climatic change. Flooding and subsequent over-bank sediment deposition will not occur if rainfall runoff fails to cross the threshold of criticality. Under these nondeposition conditions the pedogenically active surface soil may have sufficient time to weld onto a lower soil package as the front of pedogenesis reorganizes the now-fossil-buried soil.

Figure 6.1 Soil profile at Paso Otero 5 (photo courtesy of Gustavo Martinez).

The stratigraphic complexity becomes more problematic when these stable landforms are used as habitation sites by people in the past.

Paso Otero 5, an archaeological site within a complex fluvial soil, is located in the Wet Pampas region in Provincia Pampeana, south of Buenos Aires in Argentina [257]. The general stratigraphic sequence at this site consists of two members of the Lujan Formation: the late-Pleistocene Guerrero Member, containing aeolian and alluvial facies, and the early to middle Holocene Rio Salado Member, an aggrading floodplain. Recent aeolian sediments of the La Postrera cap the Lujan Formation [258] (Figure 6.1).

The primary objective of this study was to establish chronometric control throughout the soil profile, providing a context for the reconstruction of the paleoenvironment for this part of Argentina. Charcoal suitable for ^{14}C analyses was not available, and faunal material associated with the Late Pleistocene/Early Holocene period occupation had been heavily affected by biological and taphonomical processes. ^{14}C analyses of the 11 bone samples submitted returned only two credible dates that needed to be independently corroborated [258].

Paired soil samples from the upper 5 to 10 cm of selected buried A horizons were independently submitted for radiocarbon analysis to the Desert Research Institute, Las Vegas, Nevada, in independent blind studies. Dates were obtained on soil humates and on soil residue (humin fraction) for each paired sample. Two additional radiocarbon dates were obtained from bone artifacts from the Late Pleistocene/Early Holocene cultural level, designated Ab6 [258]. The rate of soil aggradation based on the soil humates and residue ^{14}C samples raised certain doubts about the presumed credibility of the bone collagen-derived dates, suggesting they should be older. The visible soil horizons and the ^{14}C dates derived from soil humates and residue suggested soil buildup and development consistent with climatic stability.

OCR analyses were conducted to independently assess which argument was most likely: that the bone collagen-derived dates were credible, or that the estimated date based on soil aggradation rates shed legitimate doubt on that credibility. The dates obtained from the OCR calculations corresponded well with those obtained through radiometric analyses for both the humate samples and the two credible bone samples, suggesting an alternative answer.

Figure 6.2 displays the visible stratigraphic column, geological formations and their members, location of archaeological deposits, and soil radiocarbon ages. Seven individual soil sola (A and B horizons) are identified. Figure 6.3 displays the OCR data for the same column and is broken into 15 pedogenic packages. Some of these event packages may indicate additional flood-buried soil levels not visible in the field; however, most are likely moderate (second-order) perturbations that have interrupted the pedogenic process. The division into packages was made following the concurrent sequencing of OCR ratio values and textures. Pedogenic horizon and corrected radiometric data are provided for comparison.

The OCR ratio generally increases in value within each separated package, indicating the increasing diagenesis to lignite of the soil organic carbon. In some of the packages, for example the field-designated A2-Cumulic package, the OCR ratio value decreases near the bottom. This decrease indicates the limit of the pedogenic front for that particular package. A new package is defined at that point where the OCR ratio begins its pattern of successive increases. The lower OCR ratio value reflects the inherited value of the non-pedogenic depositional sediment.

In some cases, for example field-designated Ab3 and Ab4, there is no drop in the OCR ratio value between soil packages. The Ab3 differs visually and texturally from the Ab4, and therefore the two still represent independent event packages. The lack of a drop in the OCR ratio value between these two definable packages indicates that the pedogenic front of the overlying package had reached and is beginning to reorganize (weld onto) the lower soil package. The lower unit may be thought of as the soil fossil undergoing taphonomical change.

The aggregate OCR ratio value at a larger scale increases with each successive defined soil package down to two levels below the field-designated Ab6 event package. The bottom two packages suddenly show a resumption of an increase in aggregate OCR ratio values, though at lower levels. This break in the sequencing of the OCR ratio correlates chronologically with the major worldwide changes in temperature and moisture regimes referred to as the Younger Dryas period.

The OCR analysis revealed greater complexity in the soil column than that assumed from the seven visible soil sola. The more than twice the number of soil packages demonstrated that the actual rate of soil aggradation did not fit a curve function from which reasonable age estimates could be extrapolated. The development of visual soil horizons masked the physiological evidence of episodic solum truncation and non-pedogenic deposition. While the visible soil horizons and

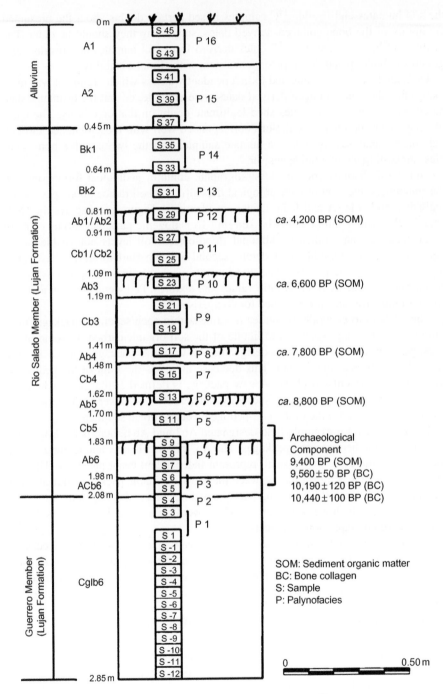

Figure 6.2 Descriptive soil profile at Paso Otero 5 with associated ^{14}C dates (courtesy of Gustavo Martinez).

Applications of OCR Dating

Soil Depth	pH	% Organic Carbon (LOI)	OCR Date*	Very Coarse	Coarse	Medium	Fine	Very Fine	Coarse Silt	Fine Silt	Sample Id	% Oxidizable Carbon (WB)	OCR Ratio	Mn	Level
6	6.9	6.308	320	0.371	0.114	0.191	0.499	5.364	57.278	36.183	6697	2.24	2.82	5.40	A1
13	6.9	6.609	380	0.158	0.107	0.223	1.121	6.382	55.100	36.910	6698	2.36	2.80	5.23	
20	7.0	5.897	822	1.501	0.173	0.355	0.733	5.232	54.581	37.425	6699	2.07	2.85	17.95	A2
27	7.0	5.742	1290	0.324	0.298	0.402	0.865	6.054	55.248	36.809	6700	1.96	2.93	13.95	
34	7.0	7.575	1776	0	0.113	0.133	0.217	3.714	48.630	47.193	6701	2.32	3.27	6.30	
41	7.1	7.859	1807	0.020	0.091	0.178	0.420	3.298	43.178	52.815	6702	5.35	1.47	11.55	
48	7.2	7.128	1870	0.078	0.092	0.299	0.621	4.681	37.146	57.083	6703	2.53	2.82	28.30	Ab
55	7.3	4.318	1958	0.066	0.128	0.353	0.730	1.860	47.642	49.222	6704	1.37	3.16	71.00	
62	7.4	3.729	2704	0.009	0.090	0.340	0.597	2.389	34.264	62.311	6705	1.11	3.36	115.00	
69	7.4	3.224	3622	0.026	0.166	0.540	0.501	1.024	28.295	69.447	6706	0.86	3.75	116.50	
76	7.4	3.049	4595	0	0.012	0.435	0.309	1.007	31.301	66.936	6707	0.79	3.86	122.00	
83.5	7.5	2.724	4738	0.014	0.017	0.418	0.397	1.212	26.791	71.150	6708	0.73	3.73	142.50	Ab2
90	7.4	2.654	5854	0.011	0.111	0.419	0.323	1.019	38.562	59.554	6709	0.62	4.28	155.00	
97	7.5	2.463	5897	0.273	0.304	0.671	0.383	0.681	35.512	62.177	6710	0.66	3.73	168.00	Ab3
104.5	7.5	2.641	6049	0.052	0.192	0.649	0.545	0.856	29.828	67.879	6711	0.66	4.00	183.50	
112	7.6	2.745	7380	0	0.022	0.681	0.281	1.152	21.687	76.176	6712	0.64	4.29	481.50	
119	7.5	2.442	7544	0.060	0.288	0.501	1.212	7.615	18.307	72.017	6713	0.58	4.21	184.00	
126	7.5	2.495	7886	0.022	0.130	0.159	0.773	2.436	36.089	60.391	6714	0.56	4.50	493.50	
133	7.5	2.454	8068	0.158	0.199	0.270	0.861	1.617	33.785	63.110	6715	0.52	4.72	196.00	
140	7.5	2.448	8329	0.281	0.299	0.486	0.733	2.260	37.977	57.964	6716	0.48	5.10	158.00	Ab4
147	7.5	1.825	8537	0.012	0.094	0.344	0.250	1.807	37.928	59.565	6717	0.39	4.68	153.50	
154	7.6	1.741	9128	0	0.118	0.340	0.336	2.142	41.111	55.953	6718	0.46	3.78	152.00	Ab5

Figure 6.3 OCR data for Paso Otero 5.

Soil Depth	pH	% Organic Carbon (LOI)	OCR Date*	Very Coarse	Coarse	Medium	Fine	Very Fine	Coarse Silt	Fine Silt	Sample Id	% Oxidizable Carbon (WB)	OCR Ratio	Mn	Level
161	7.6	1.849	9924	0.094	0.114	0.183	0.463	2.185	48.496	48.465	6719	0.34	5.44	139.00	
168	7.6	1.820	10150	0	0.114	0.460	0.742	1.707	40.128	56.849	6720	0.35	5.20	168.00	
175	7.7	1.768	10331	4.963	0.360	0.593	1.116	2.786	30.182	59.998	6721	0.39	4.53	210.50	Ab6
182	7.7	1.747	10621	0.835	0.228	0.253	0.564	3.164	38.258	56.697	6722	0.26	6.72	233.50	
189	7.8	2.151	10575	2.939	0.362	0.444	0.546	3.713	41.158	50.837	6723	0.32	6.72	278.00	
196	8.0	1.637	10572	4.521	0.556	1.240	1.324	11.057	27.654	53.649	6724	0.26	6.30	217.00	
203	8.0	1.374	10719	11.558	1.122	1.598	2.465	6.967	26.095	50.196	6725	0.35	3.93	164.00	ACb6
210	8.0	1.289	10955	4.752	0.385	0.409	0.914	8.517	41.624	43.400	6726	0.23	5.60	158.50	
217	8.0	1.239	11342	1.357	0.419	0.525	0.683	15.484	47.107	34.425	6727	0.14	8.85	143.00	
224	7.9	1.120	11037	0.007	0.118	0.135	0.399	13.924	51.351	34.066	6728	0.65	1.72	136.00	Cgb6
231	7.9	0.491	11102	1.187	0.175	0.133	0.322	17.299	55.122	25.762	6729	0.23	2.18	72.50	
247	7.9	0.561	11577	3.580	0.460	0.373	0.758	19.756	55.968	19.106	6730	0.14	4.01	30.65	
254	8.0	0.627	11893	0.968	0.286	0.366	0.695	24.280	54.546	18.860	6731	0.07	8.96	14.30	
261	8.1	0.766	12065	12.538	1.077	0.510	0.560	5.291	50.155	29.870	6732	0.13	5.89	46.75	
268	8.2	0.647	12169	0	0.131	0.156	0.217	11.939	66.294	21.263	6733	0.23	2.81	10.05	
275	8.1	0.787	12222	0.298	0.354	0.202	0.710	6.412	73.774	18.249	6734	0.17	4.63	36.30	
282	8.0	0.743	12198	0.044	0.196	0.255	0.310	7.589	64.930	26.677	6735	0.20	3.72	6.15	
289	7.8	0.850	12226	0.327	1.881	1.238	0.590	6.515	69.268	20.180	6736	0.17	5.00	1.51	

Figure 6.3 (Continued)

the ^{14}C dates derived from soil humates and residue suggested soil buildup and development consistent with climatic stability, the physiologically described soil packages along with the OCR ages suggested climatic instability during which time events of erosion and excessive deposition were interspersed with periods of no overbank flow from the river.

The discovery that climatic instability might have predominated during the Holocene period led to the next case study, which compared the growth and development of various landforms at regional and global scales.

A Look at Landforms at Regional and Global Scales

The data from Paso Otero 5 were examined along with data from four sand-dune sites along with one lagoon-edge site in the Dry Pampas north of the Rio Colorado in Provincia del Espinal, and another stream-bank site in the Wet Pampas region in Provincia Pampeana south of Buenos Aires, Argentina. This study was conducted to see if small-scale climatic variations reflected in soil physiology can be correlated across a region.

The data were entered into a "parallel coordinate plot" (PCP) [259,260], which displays an unlimited number of variables and observations by making the axes of the plot parallel rather than orthogonal, as in traditional scatter plots (in the PCP, observations are represented not as points but as unbroken series of line segments connecting the axes). This format was selected as it allows a high level of interactivity and a means of visualizing complex patterns and relationships in the data. Figure 6.4 presents a composite GIS visualization showing geographical locations at various scales, and sequential PCP presentation of data undertaken in the analysis. First, all data are displayed, followed by three separate PCPs indicating the textural patterns unique to the three individual landforms. Second, three PCPs relate pedogenic soil levels with mean age based on natural breaks. Next, these three PCPs are merged to produce one graph indicating mean age of successive climatic variation over the past 3,500 years, evidenced in the alteration between non-pedogenic sediments and pedogenic soil packages.

The alteration of non-pedogenic and pedogenic soils within these three landforms can be viewed as a complex of perturbations [33,37,253,255,256,261,262]. Recurrent perturbations of the kinds that are structurally coupled with the environment and that support organization of soil as a system are of two kinds: those that lead to the development of horizonation and those that are perceived to interrupt pedogenesis (haploidization). The latter are expressed by packages of spatially and temporally dependent, pedogenically related soils. Each package is bounded by perturbational events that define its beginning and end, exemplified by the resumption of concurrent textural and chemical sorting.

Rates and pathways of organic carbon diagenesis will differ within event-related soil packages and cannot be directly compared to those found in adjacent soil packages even within the same soil profile. Thus, it is not necessarily the case that inert carbon will predominate at lower levels while labile carbon predominates higher up in the profile. The complexes of organic carbon materials are expected to vary throughout the soil profile, depending on the unique pedogenic history of the soil.

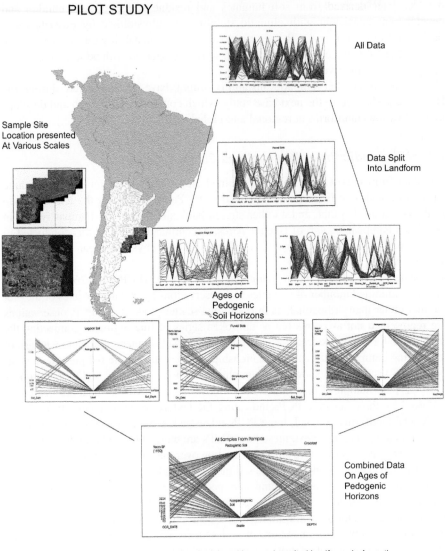

Figure 6.4 Composite GIS visualization of PCP data from Argentina.

This study was conducted as part of paleoclimatic studies for the general Pampa region (Frink, personal notes). I unexpectedly found, however, that specific landforms can also be distinguished based on the characterization of their total organic carbon and percent oxidizable content, and the quality of that carbon as expressed by the OCR ratio. The data from this study were then compared to four paleo-dune sites and four levee sites in North America (Figures 6.5–6.8).

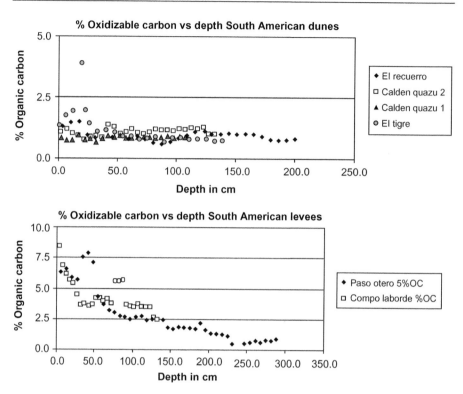

Figure 6.5 The percent organic carbon decreases with depth for both the levee and dune samples in the South American samples. The littoral soils contain considerably more organic carbon, nearly 4 times as much, especially in the upper levels, and the drop-off in amount of organic carbon is much greater than that evident in the dune soil profiles.

Much of the carbon found in levee soils comes from older carbon deposited from eroded upstream sources [263]. I expected that levee soils would reflect this bias of additional older, recalcitrant carbon (i.e., more resistant to oxidation). Levee soils are also generally finer textured than dune soils. Labile carbon is more closely associated with coarse textures, while recalcitrant carbon is more closely associated with fine textures [264–269]. All things being equal, fluvial soils should show a higher OCR ratio than dune soils.

The relative age of the samples further countered expected results based on earlier studies. The North American dunes and levee soils both cover the time span between roughly 2,000 to 11,000 YBP, but the South American levee soils are both in excess of 12,000 years old. The time-transgressive process of becoming lignite transforms organic carbon into inert organic carbon (lignite), with the organic carbon first becoming recalcitrant [230,268,270]. I did not find higher OCR ratio values for the older fluvial soils, however. The relative ages between dunes and levees were reversed for the North American samples, with no discernable affect on the OCR ratio values characterizing the two landforms.

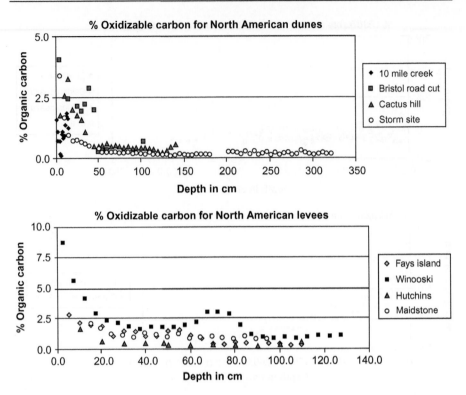

Figure 6.6 A similar, although less intensive, pattern is evident for the North American levees and dunes, with levee soils containing more total organic carbon, especially in the upper portions of the profiles, than the dunes.

The cause for the discrepancy between expected and observed organic carbon quality seen in these two landforms rests in a combination of landscape formational processes and *in situ* pedogenic processes. Much of the carbon found in levee soils comes from older carbon deposited from eroded upstream sources. A similar process may be taking place with dune formation, where older recalcitrant carbon is embedded along with sediments [271]. If both landforms inherit upstream (wind or water) older carbon macerals of similar age and makeup, I anticipate that the qualitative difference evidenced by the OCR ratio is due to the difference in degrees of enrichment between the two soils. The levee received more of the young labile, organic carbon as a result of enhanced growing conditions afforded by the adjacent available water. Recent studies on the age of dissolved organic carbon in the Amazon River tend to support my hypothesis of young carbon loading [272].

Alternatively, greater biosynthesis of organic carbon by mycorrhizal fungi on and within dunes may contribute to the greater OCR ratio values [273,274]. Mycorrhizal activity within river-bank landforms would be minimized due to frequent flooding, during which time fungitoxic chemicals are common byproducts of anaerobic decomposition of organic matter [275]. Additionally, erosion and deposition

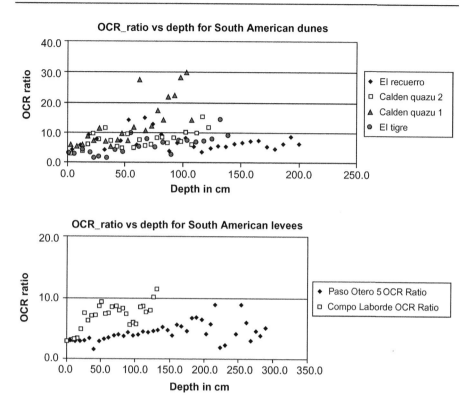

Figure 6.7 The OCR ratio value generally corresponds with depth for the levee and dune soils in the South American samples. The dune soils have nearly twice the ratio values, especially at lower levels, than the littoral soils. The higher OCR ratio suggests that more of the older organic carbon is stored in the dune soils than in the littoral soils.

perturbation sequences would impair mycorrhizal development. Mycorrhizal activity is more common in the less-perturbated periodically stable dunes.

In situ pedogenic processes affecting the quality of soil organic carbon include the biotic and abiotic processes of biochemical cycling of soil organic carbon discussed above. The finer-textured levee soils would favor abiotic processes of adsorbed and accumulated humic monomers on clay surfaces, polymerization to form humic substances with clay acting as a catalyst, and finally physically sequestering (occluding) and protecting these humic substances from further microbial action [118,119,276,277]. The dune landforms, which exhibit greater stability over time compared to the episodic aggradations of levees, favor biotic processes of microbial biosynthesis of highly aliphatic cell walls, which are chemically and physically resistant to biodegradation [124]. The abiotic process of physically sequestering organic carbon is potentially less permanent than biotic processes of biosynthesis where the remains, even upon subsequent exposure, are less susceptible to further oxidation.

The total organic carbon contained in fluvial soils is more heavily influenced by labile carbon than are the dunes, when the range of OCR values is compared.

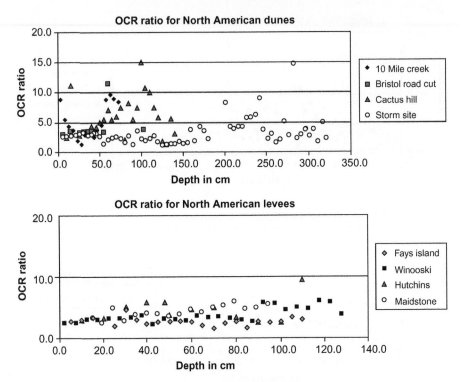

Figure 6.8 These same general patterns are exhibited for the North American samples, with dune formations evidencing higher OCR ratio values than the levee soils. The higher OCR ratio suggests that more of the older organic carbon is stored in the dune soils than in the littoral soils.

This suggests that burial of carbon as a sequestering process is dependent on the process by which burial occurs, not just geographical region, and that any predictive model of soil organic carbon pools must include the unique depositional and pedogenic history of the landform.

This hypothesis is strengthened by data from the profile obtained at the edge of a lagoon adjacent to the South American dunes (Figure 6.9). The quantity and quality of the soil organic carbon follow patterns more similar to those seen in the levee soil in both North and South America, even though the lagoon sample is located geographically near the sampled dunes.

Greater-than-expected climatic instability throughout the Holocene period, as evident in the soil physiology, was corroborated both regionally and globally. However, unexpected OCR values led to an important insight that must be taken into account when comparing soils that form on different landscape features. Different landforms create unique initial conditions for soil development processes. This affects the rates and pathways of metabolic and likely other physiological processes. Soils forming on similar landforms appear to manifest similar physiological processes. This study demonstrated

Applications of OCR Dating

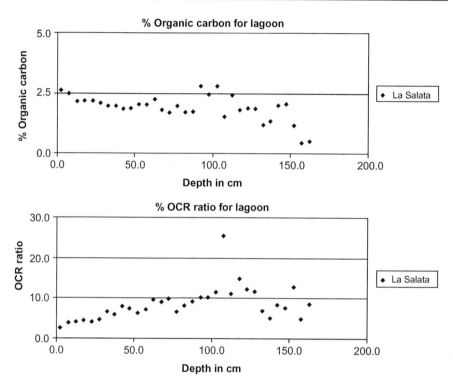

Figure 6.9 Total percent organic carbon quantitative and OCR ratio qualitative characteristics of a lagoon-edge soil follow patterns more similar to those found for the levee soils in both North and South America than for the nearby sand dune soils.

that current soil taxonomic units, based in large degree on relief, parent material, and biota, can be characterized by how the soil metabolizes organic carbon. The assessment of other soil constituents involved in soil metabolic processes will likely provide additional discriminating evidence linked to specific geomorphic landforms.

Unique Formation of Pimple (Mima) Mounds

Numerous small mound groups occurring in densities of 6 to 16 per hectare are common throughout the western-southern Mississippi River Valley area from Missouri to eastern Texas and Oklahoma, with similar mounds found elsewhere around the world, including South Africa and the western Pampas and southern Patagonia regions of Argentina. Their genesis has been the source of much debate. These mounds are similar in size, generally measuring less than 15 m in diameter, with relief of 1 to 3 m. They form in fluvial silts and clays overlying rigid, often gravelly, substrate such as former river beds. The soils that form on these pimple mounds are generally better drained and have thicker A horizons, higher cation exchange capacity and phosphorus, and lower bulk density than inter-mound soils [74].

These mounds were commonly selected over the wetter surrounding soil for past human settlements because of their drainage characteristics. The presence of human artifacts on or in these mounds has led some to believe that the mounds are human constructs. Pimple mounds look similar to indigenous house platform mounds in the region, but not all pimple mounds show evidence of human occupation. Animal burrowing and tree growth are greater on the mounds rather than between mounds for similar reasons, leading to the belief by some that these mounds are the result of this increase in biotic action [74,76,78,278–280].

Some hypothesize that pimple mounds are the result of solifluction (the action of flow in a saturated soil) of the coarser-textured subsoils due to the relationship between the surface soils and soil substrates and their geographical proximity to now or formerly seismically active areas [72,281]. Cox argues against Berg's theory, for a seismic origin for pimple mounds contending that the most extensive areas of mounded topography are located in the Rocky Mountains, Ozark-Ouchita, and Gulf Coastal Plain regions that today experience only minor seismic activity, and in the Texas Gulf Coast, where seismic activity is negligible [77]. This might be the case at present, but the effect of glacial-isostatic deformation during the late Pleistocene at the full extent of glaciation would have brought these regions into the fore-bulge area of land deformation [282]. The glaciers' retreat and the subsequent isostatic rebound would likely have resulted in significant seismic activity well into the mid-Holocene period.

It is likely that all of these factors of genesis may come into play depending on the particular mound group, and while these mounds may look alike visually, specific formational processes will be reflected in differing characteristics, a better understanding of which may lead to more refined classifications for this landform. Evaluation of the soil and archaeological interpretations for any cultural materials found on or in these mounds require the correct identification of just how the particular mound was formed.

The OCR data from archaeological site 34-MC-517 (Figure 6.10), situated on a knoll on the floodplain of Goodwater Creek, Oklahoma, demonstrate the structural complexity of these pimple mounds. The bulk of cultural material comes from the upper 35 cm, with some scattered material found to depths of between 65 and 75 cm below the surface. The recovery of temporally diagnostic ceramic shards suggests the site belongs to the Caddoan people and dates to between AD 1100 and 1400 [283].

Several lines of data are unusual relative to the typical soil profiles presented so far. First, the pH values abnormally decrease from the surface down to roughly 50 cm, at which point the trend is a slight increase in value. Second, the manganese values are an order of magnitude higher between 40 and 80 cm compared with the rest of the profile. The third aberration is found in the lack of normal patterning in most of the OCR ratio values, with the noted exception of those samples from between 65 and 80 cm. And finally, the textures are highly variable, abruptly trending both finer and coarser as one moves up through the profile.

These characteristics are not consistent with anthropogenic mounds where a sub-mound buried surface is overlain by progressively older sediments. The ages of the fill sediments are inherited from their original stratigraphic position within the borrow area such that the A horizon is excavated from borrow and placed on the mound

Soil Depth	pH	% Organic Carbon (LOI)	OCR Date	Very Coarse	Coarse	Medium	Fine	Very Fine	Coarse Silt	Fine Silt	Sample Id	% Oxidizable Carbon (WB)	OCR Ratio	Mn
5	4.0	2.475	339	1.560	0.174	0.404	12.022	31.238	20.117	34.485	4872	0.81	3.06	6.97
10	3.6	1.730	465	2.580	0.197	0.357	11.975	31.097	19.276	35.313	4873	0.63	2.77	8.03
15	3.9	1.404	499	0.598	0.120	0.309	11.975	31.097	19.990	35.910	4874	0.47	2.99	5.86
20	3.5	1.253	555	0.284	0.093	0.346	12.929	31.817	25.282	29.249	4875	0.36	3.48	6.76
25	3.3	1.101	548	0.192	0.158	0.357	12.900	33.583	19.340	33.470	4876	0.36	3.06	8.06
30	3.2	1.143	639	0.371	0.176	0.408	15.089	39.433	19.999	24.525	4877	0.26	4.40	6.55
35	3.3	1.027	631	0.821	0.120	0.359	12.397	33.163	22.030	31.110	4878	0.26	3.95	0
40	3.2	1.061	963	1.870	0.148	0.431	17.472	40.762	18.783	20.535	4879	0.23	4.61	45.69
45	3.2	0.959	905	0.504	0.323	0.430	15.042	43.889	14.729	25.082	4880	0.27	3.55	42.4
50	3.1	0.779	903	0.589	0.185	0.348	13.973	36.829	17.565	30.512	4881	0.26	3.00	39.73
55	3.2	0.875	906	0.277	0.253	0.360	18.993	34.083	20.589	25.446	4882	0.26	3.37	43.27
60	3.2	0.834	899	0.531	0.218	0.444	17.117	37.871	16.953	26.866	4883	0.26	3.21	43.46
65	3.3	0.771	943	7.803	0.152	0.569	14.470	32.737	15.913	28.356	4884	0.20	3.86	48.00
70	3.3	0.724	1028	5.615	0.159	0.421	15.266	33.736	19.798	25.006	4885	0.16	4.53	43.19
75	3.2	0.685	1106	0.330	0.152	0.362	17.508	35.748	19.891	26.010	4886	0.14	4.89	35.94
80	3.3	0.634	1175	0.219	0.146	0.316	14.601	35.816	18.800	30.101	4887	0.12	5.28	34.18
85	3.2	0.558	1188	1.401	0.147	0.311	14.857	35.062	18.789	29.433	4888	0.11	5.07	24.08
90	3.3	0.447	1562	17.856	0.116	0.266	11.300	30.374	15.283	24.805	4889	0.11	4.06	17.97
95	3.4	0.521	2485	9.937	0.100	0.298	13.443	33.373	18.452	24.397	4890	0.04	13.03	8.92
100	3.4	0.689	1803	0	0.084	0.261	12.701	35.871	17.561	33.521	4891	0.11	6.26	4.01

Figure 6.10 OCR data from archaeological site 34-MC-517.

first, with successively deeper (older) soils being added on top. Additionally, the characteristics presented above do not conform to those of bioperturbated biomantles, which are characterized by a relatively homogenized, over-thickened A horizon (similar to an Ap horizon). The age of the biomantle should increase with depth due to incremental accretion at the surface; however, this would be evident not as a definable event package but as incremental change within this specific package.

The data presented here are more consistent with seismic origin, where saturated sandy subsoils are periodically forced up through fine-textured over-burden soils as sand blows during earthquake activity. These periods of punctuated disequilibrium transport coarse sediments but little or no carbon up through the sub-mound paleosol, forming depositional packages that then undergo pedogenesis. The lack of inherited older organic carbon is evident in the similarity in age for the non-pedogenic sediments and the pedogenically active levels.

Artifacts recovered from this mound can be viewed as having a primary depositional context only within the unique soil package located at depths of between 65 and 75 cm below the surface, based on the OCR data. Those artifacts recovered from the current surface sediments are in a secondary context. This conclusion is different from the one that would be made for an anthropogenic mound where artifacts recovered within the upper sediment packages, like much of the organic carbon, are in their primary context; those from lower down would be associated with the mound fill soils, or the result of illuviation or perturbation. This conclusion is also different from an interpretation one would make for bioperturbated soils, where the stratigraphic relationship between all artifacts would be lost due to the homogenization of the soil.

Consideration of the genesis and physiological characteristics of a given landform must therefore precede archaeological interpretations. Such consideration needs to be presented as testable hypotheses and must include a variety of plausible alternatives. Diagnostic tools like the OCR procedure are invaluable resources for testing such hypotheses, as the combined data display patterns of the relationships between constituent attributes: the soil's structure. Evident changes in these patterns are indications of past second-order perturbations.

Misinterpreting Land Formational Processes

The chemical analyses used in the OCR procedure are replicable and precise, but not all interpretations of the results are guaranteed to be correct. This is true of any method. If data are modeled according to false assumptions, then the resulting age of the sample will likely be spurious. For this reason, it is always important whenever possible to gather corroborating data about the age, or other aspects, of the soil being analyzed from independent lines of reasoning. A column sample from the Puesto La Chacra site in Santa Cruz Province, Argentina, illustrates this point (Figure 6.11).

Field observations on this soil profile describe a weakly developed A horizon overlying a well-developed paleosol that extends over 1 m in depth. Extinct Pleistocene faunal remains were recovered from the lower B2t horizon at 85 to 125 cm below the surface. Figure 6.12 displays the OCR results based on the field description of the soil profile. An age of $1,067 \pm 32$ YBP for the level containing the extinct Pleistocene

Figure 6.11 Soil profile from Puesto La Chacra (photo courtesy of Ramiro Barberena).

faunal remains resulted, representing an obvious spurious date. In addition, the textural analysis revealed a well-developed Bt horizon at a depth of between 25 and 85 cm overlying a second 40-cm-thick Bt horizon. An individual well-developed Bt horizon in temperate regions can take several thousand years to form. Under the sub-arid conditions of the Santa Cruz province, where soil processes are slower due to lower annual temperatures and rainfall, the resulting OCR date is clearly spurious.

A close examination of the OCR data suggests that the soil profile may be composed of several discontinuous and truncated soil horizons. For example, the break between the Bt horizon and the B2t horizon at 85 cm likely represents two temporally separate events of long-term soil development. The manganese data, in conjunction with the patterning of the OCR ratio and textural profile, suggest another discontinuity between 26.5 and 36.5 cm. Thus, the profile should be described as an incipient A horizon developing in, or on, a paleosol (2A/B, formerly the A/B horizon). The 2A/B horizon sits on top of another but truncated paleosol (3Bt, formerly the Bt horizon). The truncated 3Bt horizon itself sits on top of another truncated paleosol (4Bt, formerly B2t horizon). At least four separate and incomplete soil packages are evident in this profile rather than the simple soil profile assumed to have developed without major perturbations. Visually these soil packages are conflated together, much like Virgil's Hippogryph, a creature composed of the hindquarters of a mighty black stag and the head, wings, and forelegs of a raven.

Soil Depth	pH	% Organic Carbon (LOI)	OCR Date	Very Coarse	Coarse	Medium	Fine	Very Fine	Coarse Silt	Fine Silt	Sample Id	% Oxidizable Carbon (WB)	OCR Ratio	Mn
8.5	5.2	8.674	20	0	0.222	0.791	17.007	52.201	22.846	6.934	7162	2.95	2.94	11.40
16.5	5.9	12.809	72	0	0.083	1.406	17.214	37.706	31.299	12.293	7163	4.74	2.70	4.30
21.5	6.3	12.683	102	0	0.076	0.692	27.903	33.590	26.855	10.884	7164	4.64	2.73	1.97
26.5	6.8	12.161	141	0	0.045	1.104	24.711	34.371	28.273	11.496	7165	3.93	3.09	1.47
36.5	7.3	11.294	180	0	0.083	1.234	30.033	34.911	22.712	11.027	7166	3.41	3.31	2.39
46.5	7.6	10.146	255	0	0.125	1.912	28.655	35.584	22.703	11.022	7167	3.05	3.33	2.20
56.5	7.5	9.056	338	0	0.161	2.185	36.295	32.097	18.567	10.696	7168	2.53	3.58	2.14
66.5	7.0	7.223	463	0	0.251	2.458	20.863	36.625	23.941	15.861	7169	1.68	4.30	2.67
101.5	6.8	9.002	1067	0	4.453	10.961	24.441	24.183	17.374	18.588	7170	1.28	7.03	3.52

Figure 6.12 OCR results based on the field description of the soil profile.

Soil research limited to a descriptive or anatomical assessment can lead to misinterpretations and erroneous conclusions. The addition of process-oriented or physiological assessments can provide a self-correcting aspect to the research, revealing incongruent data and errors or assumptions. Additionally, the combination of these two approaches can reveal hidden information on the nature and behavior of soils. An anatomical assessment alone constrains soil research by imbuing the object of study with a static quality. The addition of process-oriented or physiological assessments by its dynamic nature can expose the unique trajectories of a soil's development and evolution through time.

Environmental Studies

Climatic Complexity

Archaeologists have accepted in principle since the excavations at Herculaneum [284] that, with few exceptions, archaeological deposits are contained within a dynamic and changing context. Concurrent with this acceptance of the non-static nature of archaeological sites, however, has been the requirement, particularly in Cultural Resource Management (CRM) studies, to consider site integrity as a condition for archaeological significance. This has created what may be called the Pompeii Dilemma. Natural processes will physically and chemically affect archaeological material, and thus information, except under the most unusual circumstances, in such a manner that this material will not be discovered in its primary context or condition. Site integrity is defined by the degree to which archaeological materials retain or reflect their primary depositional context. Once an artifact is discarded, lost, or otherwise placed in or on soil, it leaves the cultural sphere of influence and becomes a part of the soil's dynamic system. It is at this moment that the archaeological record loses its primary context.

Soil science in general, and as it has been specifically applied to archaeology [285–289], emphasizes a descriptive rather than interpretive approach. This leaves the archaeologist without an understanding of the processes affecting archaeological material in an open system and specific methodologies for empirically measuring these effects. The archaeologist is left giving lip service to the influences of perturbations, while at the same time arguing integrity by piece-plotting artifacts as though their relative position at the time of discovery were a direct reflection of their primary depositional context.

Recently, soil micro-morphologists have conducted important research into the identification of key markers of past soil perturbations. These diagnostic markers reveal whether a sediment or cultural feature is in its primary or secondary context, and are invaluable for determining the significance of these deposits [290]. The cost of soil micro-morphological analysis has limited its use primarily to sediments, cultural features, and former dynamic soils (now effectively dead) in closed systems. Further work in this area could begin identifying markers of perturbations of dynamic (living) soils in open systems and provide a much-needed tool for a process-oriented soil science.

Figure 6.13 Soil profile from 8-SL-1181, Florida (photo courtesy of Johannes Loubser).

Archaeological site 8-SL-1181 in St. Lucy County, Florida, was evaluated by New South Associates for the US Army Corps of Engineers, Jacksonville District [291]. The purpose of this study was to determine the archaeological significance of the site. Recovered temporally diagnostic ceramic styles indicated a stratified multicomponent site, but the presence of one particular style appeared throughout most of the strata, suggesting a possible high degree of mixing. OCR analyses were conducted to determine if there were definable event horizons that correlated with the cultural–temporal components, and to which cultural–temporal context the persistent ceramic style belonged.

The site is located along the north bank of Ten Mile Creek and the soils are typically floodplain soils (Fluvaquents). The soils encountered during the excavation were highly variable across small areas, with the cultural material recovered from patchy areas that consisted of a thin A horizon over recent fill capping a buried Ap horizon (reflecting past use of this area as an orange grove) that in turn overlaid a bleached sandy horizon. A dark organically enriched level containing most of the cultural material was located below the bleached sands (Figure 6.13). Visually, and as first defined by local soil scientists, the soils appeared to be spodisols. Spodisols, formerly called podzol soils, are formed when coarse-textured highly acidic parent material is exposed to high amounts of rainfall. The highly acidic rainfall leaches organic material, clays, and sesquioxides out of the lower portions of the A horizon, leaving behind acid-washed, or bleached, quartz sand grains, and then precipitates these leached materials into the upper part of the B horizon, forming a dark-brown to orange-red sub-horizon designated a Bhs horizon (the "h" standing for humus and the "s" standing for sesquioxides).

It became evident upon laboratory analyses that, with pH values in the range of 6.0 and the presence of degraded bone fragments, these could not be spodic soils (Figure 6.14). Spodisols are defined not by the presence of the albic horizon (bleached sands), but by the underlying spodic horizon, which is a zone of illuviated

Applications of OCR Dating

Soil Depth	pH	% Organic Carbon (LOI)	OCR Date	Very Coarse	Coarse	Medium	Fine	Very Fine	Coarse Silt	Fine Silt	Sample Id	% Oxidizable Carbon (WB)	OCR Ratio	Mn	
2.5	5.4	2.06	247	0.571	1.374	11.809	54.546	18.988	9.588	3.123	6407	0.32	6.54	2.33	A
7.5	5.7	0.852	209	9.472	1.683	11.247	49.428	18.978	7.521	1.670	6408	0.21	4.06	2.00	Fill
12.5	6.2	1.22	596	0.013	1.135	11.694	57.723	18.785	7.576	3.075	6409	0.26	4.69	2.475	2Apb
17.5	6.2	1.507	780	0.010	1.017	12.802	55.105	19.364	7.518	4.185	6410	0.41	3.68	0.29	2Apb
22.5	6.2	0.32	1090	0.000	1.223	13.052	57.996	18.877	7.442	1.410	6411	0.11	2.91	0.09	2Ab/C
27.5	6.1	0.697	1277	0.145	1.474	14.453	56.723	18.364	6.843	1.997	6412	0.27	2.58	0.005	2Ab/C
32.5	6.0	1.478	1777	0.349	1.632	14.156	55.041	21.106	5.479	2.237	6413	0.37	3.99	0.005	3Ab
37.5	6.0	1.14	2435	0.221	1.714	14.444	55.313	20.564	5.664	2.079	6414	0.28	4.07	0.005	3Ab
42.5	6.0	0.852	2635	0.726	1.673	14.740	53.454	20.199	7.077	2.130	6415	0.28	3.04	0.005	3Ab
47.5	5.9	0.809	3364	0.200	1.660	14.668	54.195	20.307	6.551	2.419	6416	0.19	4.26	0.005	3Ab
52.5	5.8	0.964	3548	2.681	1.629	14.537	49.777	20.621	6.966	3.789	6417	0.33	2.92	0.005	3Ab
57.5	5.8	1.074	4198	9.317	1.295	11.826	46.326	18.845	7.759	4.634	6418	0.24	4.46	0.005	3Ab
62.5	5.9	1.299	4548	10.537	1.948	13.165	45.030	19.101	6.469	3.750	6419	0.19	6.84	0.005	3Ab
67.5	5.9	1.164	4788	6.395	1.558	13.833	47.806	19.076	7.393	3.939	6420	0.27	4.31	0.57	4Ab
72.5	5.7	0.912	5592	0.000	1.692	14.265	50.876	21.730	7.444	3.994	6421	0.19	4.80	0.53	4Ab/Bs

Figure 6.14 OCR data for archaeological site 8-SL-1181.

organic matter and sesquioxides. The leaching of carbonates and significant replacement of exchangeable cations by hydrogen and aluminum ions (acidification) in the A horizon are prerequisites to the mobilization of organic matter [57]. Analysis of the OCR data indicates seven soil packages, including the developing A horizon and recent fill for the sampled soil column. Of particular note are the OCR ratio values for the buried Ap horizon and the associated C horizon (field designated as 2Apb and 2Ab/C) and the manganese value for the 3Ab horizon.

It is common to have mixed OCR ratio values (high to low trending) within the plow-perturbated zone; however, it is not typical for these trends to continue into the sub-plow soils. The site's location adjacent to Ten Mile Creek might suggest the possibility of incremental flooding, but this would not account for the atypically high to low trending OCR ratio values.

This unique soil morphology can be understood by imagining what a profile might look like if a hole was dug in 10-cm levels and the excavated contents were carefully placed on top of the previously excavated level. The dirt pile would then exhibit a stratigraphy inverted from that encountered in our original hole. One cannot know for certain the event or events that resulted in the inverted soil data encountered at this site, but testable hypotheses can be posed drawn from the OCR data.

If the OCR ratio inversion were the result of cultural processes such as mound building, a correlated increase in organic carbon with depth is expected, and such is not the case. If the OCR ratio inversion were the result of natural deposition such as aeolian processes, homogeneity along with a lower percentage of organic carbon values when compared to adjacent soil packages is expected.

Paleoclimatic data from New Mexico, Tennessee, Vermont, and the north-central Yucatan Peninsula, Mexico, all indicate a dramatic shift in climatic regimes after roughly 1,800 YBP. Analyses of columnar stalagmites in New Mexico show a shift from wetter to dryer conditions at $1,749 \pm 38$ YBP. A concurrent cultural change corroborating these conditions is seen in the shift from corn- to cotton-based agriculture [292]. Stable oxygen isotope values of sagittal otoliths from an archaeological site in northeastern Tennessee reveal colder conditions before 1,800 YBP giving way to warmer conditions by 1,700 YBP [293]. Hydrologic records from lacustrine sediments in Vermont indicate a change from high storm runoff prior to 1,750 YBP to lower runoff values since [294]. Lake sediment cores from the Yucatan Peninsula similarly indicate a dry period between 1,825 and 1,740 calibrated YBP, followed by a return to wetter conditions [295]. Again, a major change in cultural patterns between the demise of the Maya Preclassic period to the rise of the Classic period correlates with this change from dry to wetter weather patterns [296]. The extent of this climatic shift is also indicated throughout much of South and Central America as well as the western USA [297].

The OCR data from Paso Otero 5 presented above (Figure 6.3) also show evidence of a second-order perturbation at this time period (1,776 YBP) that may be climate related. Moisture regimes in South and Central America have been found to vary inversely with those of the southeastern USA [298]. Concurrent evidence for a climatic shift throughout the Western Hemisphere supports the argument for a climate-induced second- or even third-order perturbation affecting the soil at archaeological site 8-SL-1181. Taken as a whole, these studies suggest that the various weather-pattern shifts would have translated into dryer conditions in the Florida peninsula during the second half of the second century AD.

The manganese values, although low in all soil packages, are undetectable throughout the organically enriched 3Ab horizon, where the bulk of the cultural material was recovered. A first thought on this matter might logically be that the manganese, being water soluble under reduced conditions, may have leached from the soil column. The presence of manganese in the soil package 4Ab and 4Abs horizons that underlie the 3Ab horizon, and in the 2Apb and 2Ab/C horizons above the 3Ab horizon, however, argues against this explanation. The absence of detectable available manganese in the 3Ab horizon instead suggests that its soil package is unique and different from the pedological processes affecting the other soil packages.

The 3Ab horizon is visually one continuous horizon, but based on the aggregate data the 3Ab horizon can be segregated into three temporally and spatially separate subunits. The pedogenic ages of these three subunits correspond well with the typological ages of the ceramic styles recovered from this horizon, as well as with associated radiocarbon ages, all dating before 1,700YBP (Figure 6.15).

The best explanation for the 3Ab horizon is that it is an aggraded sheet midden, composed of the detrital accumulations of life's activities at camp: ash from fires, food wastes, and debris from tool manufacture and repair. Some mixing of coarse particles such as artifacts would be expected during each successive occupation [299], but the separate subunits definable from the OCR data suggest that such mixing was limited. The presence of the ceramic style found throughout the 3Ab horizon midden is likely a style that persisted through all the represented cultural components.

The best hypothesis to explain the bleached sand deposit overlying the midden horizon that dates to sometime soon after 1,777YBP, a period when abnormally dry weather patterns for this area are indicated by large-scale paleoclimatic reconstructions, is that this horizon likely represents a small sand-dune formation or sand sheet. Soil moisture and pH values are higher in these areas because of the enriched midden deposits, and these conditions would have allowed vegetation to survive the drought conditions better than those on nearby soils. This stand of surviving vegetation would capture the aeolian sands as wind erosion depleted nearby unvegetated landforms, filtering them out of the air as well as protecting them from becoming re-entrained by the wind. Dunes that form through this process are known as coppice dunes.

Paleoenvironmental Reconstructions from Anthropogenic Landforms

Stress placed on human populations resulting from shifts in the weather or climatic patterns can induce changes in human exploitation patterns of the environment. Understanding the role of special-use sites in the overall settlement pattern of a culture may be determined through comparisons between climatic patterns and occupational events, as evident in the punctuated dynamics at site RB3-A, an archaeological shell mound near Weipa, Australia (Figure 6.16).

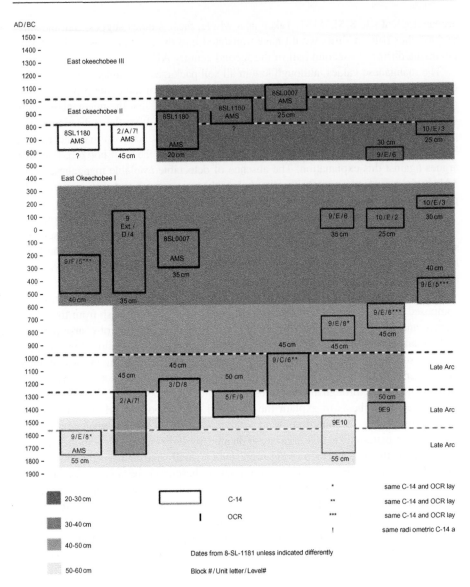

Figure 6.15 Schematic showing relationships between soil levels and both OCR dates and radiometric ages [291].

Applications of OCR Dating 65

Figure 6.16 Composite photograph of shell midden RB3-A, Weipa, Australia (photo courtesy of Michael Morrison).

OCR analyses (Figure 6.17) of each depositional event and the subsequent pedogenic weathering was found to show a strong correlation with recorded historic El Nino Southern Oscillation (ENSO) events (Figure 6.18). During these events the usually warm and moist regime of the Cape York Peninsula is disrupted by severe and persistent drought. It may be hypothesized from this correlation that drought conditions and the high potential for forest fires in the interior forests would have enticed

Soil Depth	pH	% Organic Carbon (LOI)	OCR Date	Very Coarse	Coarse	Medium	Fine	Very Fine	Coarse Silt	Fine Silt	Sample Id	% Oxidizable Carbone (WB)	OCR Ratio	Mn
9	7.0	26.387	38	67.877	3.779	1.627	3.493	2.940	4.333	15.952	6293	13.43	1.96	3.1
13	7.0	30.134	172	68.586	2.784	1.688	3.306	3.127	6.117	14.390	6294	8.33	3.62	2.0
17	7.0	34.942	187	64.099	2.859	2.101	3.558	3.541	7.747	16.095	6295	10.47	3.34	1.8
22	7.0	27.437	201	60.412	4.944	2.813	4.317	3.984	7.610	15.919	6296	9.70	2.83	1.4
27	7.0	13.723	220	58.540	6.081	3.246	5.196	4.263	7.312	15.362	6297	7.60	1.81	1.5
32.5	7.0	14.297	231	66.218	4.129	2.799	4.865	3.820	9.218	8.951	6298	8.17	1.75	1.3
40	7.0	9.836	244	70.628	4.819	3.034	4.497	4.192	5.662	7.169	6299	5.66	1.74	1.8
55	7.0	10.536	263	72.071	4.268	2.989	5.564	4.267	5.015	5.827	6300	4.17	2.53	1.75
62.5	7.0	8.213	285	46.339	7.818	5.066	11.300	9.163	5.670	14.643	6301	3.68	2.23	1.9
67.5	7.1	7.02	308	71.402	3.659	2.030	6.356	4.805	2.373	9.375	6302	2.82	2.49	2.4
80	7.1	3.008	330	81.273	4.829	1.847	2.329	1.344	1.293	7.085	6303	1.79	1.68	0.8
87.5	6.9	10.113	344	67.305	6.308	3.067	3.338	3.600	3.073	13.309	6304	5.73	1.76	1.5
92.5	7.1	3.634	364	64.545	7.033	2.546	3.938	2.868	1.432	17.638	6305	2.46	1.48	2.1
97.5	6.9	3.037	395	34.600	3.908	3.981	31.689	18.448	1.711	5.664	6306	1.66	1.83	2.32
102.5	6.8	2.043	488	9.725	2.587	4.879	38.523	40.067	2.167	2.052	6307	0.54	3.78	2.585

Figure 6.17 OCR data for Weipa shell mound.

OCR date: YBP	OCR date: AD	Recorded ENSO events (Caviedes, 2001)
38±1	1912±1	1911
172±5	1778±5	1777–1778
187±6	1763±6	1761
200±6	1749±6	1747
220±7	1730±7	1728
231±7	1719±7	1714–1715
244±7	1706±7	1707–1708
263±8	1687±8	1687–1688
285±9	1665±9	1660
308±9	1642±9	1634
330±10	1620±10	1618–1619
344±10	1606±10	1607
364±11	1586±11	1591
395±12	1555±12	1552

Figure 6.18 Comparison between OCR dates and ENSO events for the RB3-A shell mound at Weipa, Australia. (Shaded lines indicate specific events discernable from the OCR data.)

populations to move to coastal areas in search of food and safer living conditions. The shell mound is predominated (90%) by *Anadara granosa*, live cockle. Live cockles are extremely sensitive to turbidity in their breeding areas, a condition that would prevail during the usually wet conditions of this area. However, during the early ENSO drought events water temperatures in this area increase, turbidity decreases, and nutrient upwelling occurs in the surrounding oceans, establishing ideal conditions for a live cockle bloom within the estuaries surrounding the peninsula. The apparent discrepancy found between the OCR dates and independently obtained ^{14}C radiocarbon dates from shell samples obtained from the various deposit levels is of particular interest in this study. Unlike the previous case study, where the spurious OCR results led to new ideas about the soil's complex history, here the OCR results exposed some of the problems encountered with ^{14}C radiocarbon dates and the assumptions commonly made about those results.

The standard radiocarbon dates obtained from the analysis of shell carbonates must be both calibrated to calendar years and adjusted for the reservoir effect for the original source of the carbonate that varies throughout the world's oceans depending on temperature, circulation, and sample species. The general regional mean for northeastern Australia is based on three invertebrate species obtained from the Torres Strait and two invertebrate species obtained from roughly 80 km north of Weipa. The two species from north of Weipa are *Telescopium* and *Anadrara*. The data obtained from these two species represent the extremes for the regional mean. The corrected and calibrated ^{14}C results are presented based on the mean of the three species.

When corrected to just the *Anadrara* data from near Weipa, the ^{14}C dates correlate with those obtained through OCR analyses.

Environmental Perturbations in the Dzban Burial Mounds

The following case study illustrates the patterned data indicating the self-organizing processes of soil as they occurred at Mound 1, Dzban, Moravia, Czech Republic (Figure 6.19). As the mound is a constructed monument, a high degree of stratigraphic complexity was observed in the descriptive soil profiles. Generally the soils consist of a 5-cm-thick A/E surface horizon overlying a thin 2-cm-thick E horizon. An emergent stone line is contained within the Bs horizon that extends from 7 to 15 cm below the surface and overlies an incipient Bt horizon. This upper solum rests on, and is forming in, non-pedogenic mound fills that extend from 15 to 37 cm below the surface. The mound fill rests on a roughly 7-cm-thick prepared floor. The sub-mound floor was created by adding pea-sized and smaller coarse material (likely river-run sands) to the now buried paleo-surface (2Ab horizon). A 2-cm-thick relatively unaffected portion of the 2Ab horizons extends below the prepared floor. A 2Bb horizon extends from 46 to 55 cm below the surface and sits upon a well-developed 2Btb horizon that extends to the base of excavation at 60 cm. The pedogenic data obtained from the OCR analyses are similarly complex [255].

One moderate to strong second-order pedogenic perturbation is evident in the OCR data from 15 cm below the surface (see Figure 6.19). This perturbation is evident in a change in the manganese trend along with textural sorting in samples from 15 to 18 cm below the surface. An interruption in the OCR ratio trend is also evident at this point, suggesting that perturbational events are recorded in the soil as packages similar to those of fluvial soils discussed in the previous chapter. This recent perturbation of the mound dates to 1,300 ± 39 YBP. This perturbation may relate to a meteorite that is reported to have impacted the earth in AD 536, loading the atmosphere with dust and debris and affecting weather throughout Europe for several decades [300].

The paleosol located below the mound includes the prepared floor in the upper 2Ab horizon. A strong discontinuity signifying the construction of the mound exists between the upper mound fill (30 cm and above) and the lower paleosol (33 cm and below), and it is evidenced by pH, percent organic carbon, and manganese values that, although random from sample to sample, trend to lower values with depth. These data are an artifact of the now-fossilized paleosol. A scalar shift in textures is also indicated at this point. The construction of Mound 1 occurred after 5,236 ± 157 YBP. The typology of the intact ceramic cremation vessels recovered from within the mound's burial chamber corroborates this date [301–304].

A moderate second-order perturbation is expressed in the data from the sample at 48 cm below the surface. Again an interruption in the OCR ratio value trend is accompanied by a marked increase in percent organic carbon and pH values, along with textural sorting, all indicating a new soil package. The calculated OCR date for this event is 7,536 ± 226 YBP. The change from Mesolithic period hunter-gathering cultures to Neolithic period agriculturalists occurred in central Europe at about this

Applications of OCR Dating

Soil Depth	pH	% Organic Carbon (LOI)	OCR Date	Very Coarse	Coarse	Medium	Fine	Very Fine	Coarse Silt	Fine Silt	Sample Id	% Oxidizable Carbon (WB)	OCR Ratio	Mn	
3	2.9	9.629	170	14.304	3.566	1.841	1.429	2.850	14.802	61.208	6051	4.12	2.34	57.54	A/E
6	2.9	8.329	300	22.777	3.185	1.790	1.456	1.776	19.505	49.511	6052	4.07	2.05	30.36	E
9	2.9	5.246	515	33.856	4.038	1.692	1.437	18.597	7.338	33.041	6053	1.79	2.93	6.16	Bs
12	2.9	3.449	861	32.565	6.054	1.955	1.354	1.593	24.353	32.125	6054	1.00	3.45	5.41	
15	2.9	3.065	1300	23.593	6.926	3.890	1.576	5.814	26.359	31.843	6055	0.815	3.76	5.60	Bt
18	2.9	2.663	1721	18.348	8.184	3.829	3.552	1.445	11.070	53.573	6056	0.88	3.03	9.27	
21	2.8	2.434	2367	11.671	7.748	4.572	4.556	7.160	7.434	56.860	6057	0.63	3.86	10.89	
24	2.8	2.071	3100	14.074	6.992	3.879	4.314	16.717	5.958	48.067	6058	0.48	4.31	13.14	Mound fill
27	2.8	2.112	3947	11.975	7.152	6.225	7.412	14.353	8.163	44.720	6059	0.42	5.03	15.26	
30	2.8	1.994	5098	16.748	8.758	5.965	7.856	14.365	5.150	41.158	6060	0.28	7.12	19.45	
33	3.1	2.011	5236	22.793	9.667	7.487	4.460	7.193	6.950	41.450	6061	0.53	3.79	47.81	Floor
36	3.2	2.244	5282	12.633	6.592	8.390	4.696	4.813	4.813	58.062	6062	0.54	4.16	50.92	2Ab
39	3.3	2.256	5361	7.950	6.621	8.918	3.179	5.594	7.717	60.021	6063	0.43	5.25	55.67	
42	3.4	2.206	6417	6.848	6.730	3.307	4.232	5.072	18.439	55.372	6064	0.375	5.88	48.74	2Bsb
45	3.5	2.056	6732	5.988	4.614	2.563	1.524	5.646	26.744	52.921	6065	0.38	5.41	44.935	
48	3.7	2.402	7536	7.234	12.227	3.547	.436	3.099	22.386	51.071	6066	0.46	5.22	49.23	
51	3.9	2.063	8647	8.691	6.033	3.030	1.038	10.719	14.360	56.129	6067	0.32	6.45	28.35	2Btb
54	4.0	2.128	9807	10.596	6.335	7.717	14.319	14.992	10.971	35.070	6068	0.25	8.51	25.55	
57	4.0	1.538	11644	8.159	8.610	6.422	17.365	9.990	9.731	39.723	6069	0.14	10.99	19.44	

Figure 6.19 OCR data from top of Dzban mound.

time. The large-scale deforestation of this area that accompanied the cultural change may be responsible for this second-order perturbation [305]. Additional information on genetic and diagenic processes is evident along the toe-slope portions of the mound (Figure 6.20). A major discontinuity between 25 and 30 cm is evident in the OCR ratio value and in a scalar shift in textures. The manganese data, along with the OCR ratio values, indicate that pedogenic soils above the discontinuity have been in place undergoing pedogenic processes since approximately 5,123 ± 154 YBP. The non-pedogenic mound sediments rest on top of a paleosol, forming the second major discontinuity between 48 and 52 cm. Again the OCR ratio value and scalar shift in textures show this discontinuity.

The absence of any evidence of pedogenesis in the colluvial sediments suggests that the paleosol was buried during mound construction after 5,371 ± 161 YBP but before 5,123 ± 154 YBP. An interruption in the OCR ratio value trend at 70 cm, dating to 7,627 ± 229 YBP, may be the result of deforestation during the change from Mesolithic to Neolithic cultural land uses as noted in the previous profile (see Figure 6.19).

The taphonomic processes affecting this 5,250-year-old mound are numerous and complex, as reflected in the data (see Figures 6.19 and 6.20). These measurements describe conditions trending toward both horizonation and mixing. These various processes also demonstrate the processes of self-organization in soil. Textural sorting occurs in response to second-order perturbational events. Both coarse and fine soil particles are translocated downward through the soil profile, with the coarse fraction moving somewhat slower than the fine fraction, following the perturbation. Organic carbon is oxidized and similarly translocated downwards, roughly following the coarse fraction. Manganese values decrease with depth down to the same level as the depth of actively biodegraded organic carbon, at which point they increase, indicating the terminal front of ongoing pedogenesis, which is the oldest level of the current event horizon.

Paleoclimatic Reconstruction from a Natural Landform

Bristol Pond, in Bristol, Vermont, is one of several extant freshwater ponds left behind after the draining of Glacial Lake Vermont. The environmental importance of such postglacial freshwater ponds to early Native Americans has been argued from the perspective of thermal refugia [306] and as high-yield niches within a complex forest mosaic similar to those used throughout all periods of Native American history [307]. Numerous very early Native American sites are known along the edges of Bristol Pond (Figure 6.21).

A recent road-cut enhancement along Monkton Ridge Road in Bristol exposed a 3-m-deep soil profile, providing an opportunity to examine the paleo-landscape and environment during this period of first human occupation. The road cut is located adjacent to a USGS benchmark at 160 meters (msl), placing the soil profile between 160 and 163 meters (msl). All of the identified early Native American sites along Bristol Pond are at elevations below 158 meters (msl); thus, the pedogenic history available from this soil profile can shed light on the paleo-landscape at the time these sites were occupied.

Applications of OCR Dating

Soil Depth	pH	% Organic Carbon (LOI)	OCR Date	Very Coarse	Coarse	Medium	Fine	Very Fine	Coarse Silt	Fine Silt	Sample Id	% Oxidizable Carbon (WB)	OCR Ratio	Mn	
5	3.0	11.436	359	3.852	1.456	1.449	1.838	1.437	13.352	76.617	6110	5.17	2.21	71.01	A/E
10	2.9	4.365	1067	4.558	2.785	2.160	1.050	1.061	4.295	84.090	6111	1.74	2.51	15.64	E
15	2.9	2.889	2235	4.396	2.522	1.638	.975	.832	4.597	85.040	6112	0.90	3.21	5.96	
20	2.8	2.359	3569	3.941	2.487	1.734	1.042	.736	6.440	83.618	6113	0.73	3.23	4.87	Bs
25	2.9	2.149	5123	9.128	1.944	1.237	1.228	1.092	10.471	74.901	6109	0.525	4.09	5.03	
30	2.9	2.216	5283	14.134	3.778	1.613	.842	.754	6.647	72.233	6114	0.84	2.64	6.19	B/C colluvium
35	2.9	2.512	5277	13.127	5.514	2.126	1.054	.865	7.882	69.431	6115	0.90	2.79	6.87	
40	2.9	2.570	5255	26.359	4.204	2.072	.870	.633	16.737	49.124	6116	0.84	3.06	7.71	
45	2.9	2.249	5430	25.633	4.065	1.950	.941	1.327	6.835	59.250	6108	0.38	5.92	8.93	
48	2.9	2.224	5276	26.502	5.141	2.529	1.103	.967	6.486	57.273	6118	0.685	3.25	11.45	
52	2.9	2.303	5371	9.893	4.978	2.204	1.592	1.212	3.416	76.705	6119	0.61	3.78	15.3	2Ab
55	2.9	2.073	5708	11.231	4.913	2.443	1.845	1.213	3.543	74.813	6120	0.39	5.32	16.81	
60	2.9	2.098	6205	7.804	4.665	3.368	1.690	1.392	9.305	71.775	6117	0.29	7.23	20.69	2Bsb
65	3.0	2.281	6446	19.889	5.146	2.822	2.951	1.100	2.966	65.127	6121	0.45	5.07	14.76	
70	2.9	2.350	7627	20.754	4.300	1.950	2.093	1.566	3.092	66.246	6122	0.56	4.20	17.03	
75	2.9	2.231	9145	31.360	5.972	3.572	2.646	1.492	3.009	51.949	6123	0.35	6.37	20.73	
80	2.9	2.118	10653	29.345	7.647	4.637	3.732	1.894	3.457	49.288	6124	0.34	6.23	23.93	

Figure 6.20 OCR data, toe of Dzban mound.

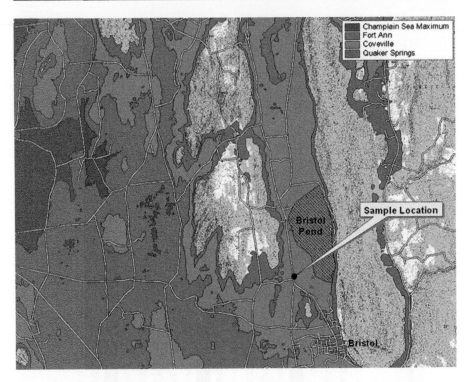

Figure 6.21 Location of Bristol Pond study area showing Glacial Lake stages and early Native American site locations (red triangles). (For interpretation of the references to color in this figure legend, the reader is referred to the Web version of this book.)

The exposed soil profile consists of an over-thickened plow zone overlying a well-developed buried soil that had formed within a paleo-dune. Another thin paleosol formed in water-deposited, bedded fine sands and silts, and lies below the dune (Figure 6.22). OCR analyses were preformed on a column sample from this profile to determine the age of this paleo-dune. The maximum age of the soil forming on top of the dune is 11,271 ± 338 YBP, and the lower paleosol below the dune deposit dates to before 12,215 ± 366 YBP (Figure 6.23). The existence of an active dune overlying littoral sediments suggests that dry and perhaps cold conditions existed in Vermont during this period.

A recent study on the style and deposition of humid-temperate fans in three Vermont towns supports the hypothesized dry conditions at this time in history [308]. Five nearby trenched and dated fans have been sampled, three of which have basal (subfan) sediments that date back to this time period: Eden at 13,320 ± 50 with a formerly pedogenic surface dating to 12,900 ± 40, Bristol at 12,980 ± 40 with a formerly pedogenic surface of the same age, and Bridgewater at 11,330 Y.B.P. without evident surface pedogenic development (dates given as calibrated ^{14}C Y. B.P. Radiocarbon analysis conducted by the Center for Accelerator Mass Spectrometry, Livermore Lawrence Laboratory, and calibrations based on CALIB version 4.2). Fluvial sediments marking

Figure 6.22 Soil profile from the Bristol Pond sand dune.

the beginning of these fans are deposited on top of the eroded paleosols [308], suggesting that the actual age of initial fan development postdates the ^{14}C ages of the two paleosols.

Evidence for a period of stability sufficient for soil development on the initial fan deposit survives only in the Eden fan deposits. This relic paleosol dates to a period prior to 9,500YBP. A second major fan-building event at both the Eden and Bristol occurred around 9,400YBP. This event is corroborated by an independent study at Ritterbush Pond, also in the Town of Eden, where a date of 9,440cal ^{14}C years YBP was obtained from lacustrine sediments marking a series of major flood events [294].

Oxygen isotope ratios obtained from six Greenland ice cores (Camp Century, Dye-3, GRIP, GISP2, Renland, and North GRIP) are reliable proxies for temperatures in the Northern Hemisphere [309]. A numerical age of 11,500YBP is deduced for the climatic change from the colder Younger Dryas to the warmer and wetter Preboreal based on annual cycles in calcium and ammonium ion records from the Greenland ice cores. This climatic change took place quickly over just a couple of decades at most [310–312].

Soil Depth	pH	% Organic Carbon (LOI)	OCR Date	Very Coarse	Coarse	Medium	Fine	Very Fine	Coarse Silt	Fine Silt	Sample Id	% Oxidizable Carbon (WB)	OCR Ratio	Mn
5	6.2	4.002	202	4.140	2.377	2.983	8.027	14.395	26.233	41.846	6317	1.40	2.86	10.85
15	5.8	2.409	1105	6.812	2.750	2.661	6.511	16.148	25.892	39.227	6318	0.705	3.42	7.8
25	5.3	2.104	1997	6.666	2.630	2.377	6.246	14.844	31.806	35.431	6319	0.65	3.24	7.0
30	5.4	1.934	2511	7.762	2.303	2.428	6.070	16.162	32.370	32.905	6320	0.58	3.33	5.2
35	5.5	2.19	2817	10.096	3.042	2.633	6.529	13.158	26.288	38.252	6321	0.62	3.53	5.8
40	5.5	2.854	3191	7.768	3.180	2.634	6.778	14.706	24.816	40.119	6322	0.80	3.57	6.25
45	5.6	1.993	3796	4.792	1.845	1.626	4.435	12.578	40.010	34.715	6323	0.545	3.66	4.22
50	5.7	0.294	8377			.080	.623	18.941	55.826	24.530	6324	0.06	4.90	1.56
55	5.6	0.306	9573			.050	1.061	29.442	50.248	19.199	6325	0.09	3.40	1.39
60	5.7	0.341	11271			.080	3.573	42.113	41.215	13.018	6326	0.03	11.37	3.225
103	5.3	0.676	12215			.057	2.937	27.367	41.094	28.545	6327	0.18	3.76	6.3

Figure 6.23 OCR data for sand dune soil profile at Bristol Pond.

Dune development at Bristol Pond may logically be hypothesized to have been active during the relatively cold and dry conditions of the Younger Dryas. A thin, buried soil formed in the upper portions of Glacial Lake Vermont littoral sediments underlies the dune, and its thinness and cambric (early) development indicate that only a brief period of time elapsed between littoral conditions and dune formation. The landscape around Bristol Pond at elevations in excess of 160 meters (msl) would have been habitable, but landscapes at elevations below 160 meters would have likely remained inundated prior to 12,000 years ago, until the waters impounded by Glacial Lake Vermont receded below this elevation. Therefore, it is reasonable to assume that lands below this elevation would not have been sufficiently dry for plant and human habitation until sometime after 11,500 years ago, when climatic conditions became similar to those experienced throughout the later Holocene period.

These data argue against the extreme antiquity (ca. 12,000 YBP or earlier) based on artifact typology alone for the Paleo-Indian period sites located around Bristol Pond and elsewhere in the Champlain Valley of Vermont [313,314]. The results from this OCR study support instead the hypothesis that migration of early people into the Champlain Valley came first by way of an Atlantic coastal route, then up the St. Lawrence River Valley, and then up the Richelieu River Valley into the Lake Champlain Valley [315].

A process-oriented soil science can greatly inform cultural understandings about the context of archaeological data. Once cultural material is discarded or lost and enters the soil, it is transformed, both physically and chemically, according to the soil's physiological processes. Distinguishing and sequencing the perturbational history of a soil is essential for understanding when and under what conditions cultural material enters and is subsequently transformed by the dynamic soil system. Clues about climate and climate change are often suggested in the soil's perturbational history, providing archaeologists with testable hypotheses on the connection between climate change, environmental change, and cultural change.

Cultural Studies

Assessment without Desecration of Sacred Spaces

Cultural complexity can mean many different things, each having its own unique impact on or relationship to soil. Here the focus is on the relationship between archaeologists and aboriginal peoples with regard to studying aboriginal sacred and special sites.

After nearly a decade of consternation and resistance to repatriation laws, archaeologists in the USA are beginning to hear, if not understand, indigenous peoples' concerns over burials and other aspects of the sacred. Similarly, this trend in sensitivity to indigenous concerns is increasing worldwide.

The Native Americans Grave Protection Act (NAGPRA), passed in 1990, requires that archaeologists consult with Native peoples concerning the treatment of burials discovered during excavations. A burial is no longer treated as just another class of artifact or archaeological feature that is subjected to clinical analysis. Rather, burials and other sacred and special sites are functional components of contemporary Native cultures. More importantly, the treatment of these sites must be undertaken under consultation with Native authorities [316]. This change in authority has led to a concurrent change in methodological approaches based on greater use of ethnohistorical information maintained through family stories and shared lessons of Native peoples.

Archaeological studies in the past were fraught with a conflict between tangible evidence from the past and seemingly less tangible histories about the past. Artifacts seem to provide immutable characteristics of truth that can be empirically measured, classified, and displayed. Histories, in contrast, are perceived as subjective and derived from interpretations. Traditionally, archaeologists chose to resolve these tensions by ignoring the histories of the cultures whose ancestors they purported to study. The archaeologists' dialogue with the past was little more than a monologue held in isolated seclusion, with findings restricted to statements about discovered things untempered by the annealing powers of interpretation.

Archaeologists' traditional obsession with artifacts (things), rather than behavior (processes), has limited the inclusion of traditional oral histories as meaningful data on a par with artifacts that can be subjected to empirical study. People and their histories, on the other hand, are seen as data that is *a priori* suspect, biased, and reconstructed. The conflict between the objectivity of science and the subjectivity of nonscience is rooted in Enlightenment-era philosophies. This conflict, however, is and always has been based on a false dichotomy [45].

The recent changing attitudes in archaeology are helping to create a long-overdue trust with Native cultures. This new trust makes possible a mutual sharing of information as well as equal access to the means and technologies for seeking answers to questions [317]. Burials do exist and are inadvertently encountered during archaeological studies as well as during construction projects. Consultation by archaeologists with Native peoples as to the meaning, disposition, and repatriation of these remains is just a first step in the proper stewardship of historical and sacred resources. New techniques must also be adapted for the documentation process as well, techniques that not only fulfill the obligations of the archaeologist as a scientist but also respect peoples' beliefs. One such technique is to focus archaeological inquiry toward anthropogenic soil perturbation, the digging and refilling of the burial pit, rather than on the physical remains of the human body itself or the material artifacts included in the burial (sacred grave goods).

Recently, the discovery of human remains in the back dirt piles of an excavated house foundation prompted a court-ordered work stoppage. A subsequent recovery effort was initiated by the authorized Native government. (The locational context is intentionally omitted in this discussion at the request of, and out of respect for, the Native peoples whose ancestral remains are the subject of this case study.) A request for assistance was made to area archaeologists to assist in this effort. In addition to the concern that these human remains be quickly and respectfully gathered and properly repatriated, the local Native government was also interested in the potential information this unfortunate event might provide for their culture. One important question was the historical context of this burial ground; they needed to know its age so that they could more easily link the burial event with family descendents, whose participation in the reburial is an important concern.

Several burial shafts were partially exposed in the excavated side walls in addition to the human remains recovered from the back dirt piles. The local Native government requested that I conduct OCR analyses on one of these exposed burial shafts to date the age of interment. They requested, however, that I conduct the study in a manner that would avoid the actual body and the entombing soils immediately surrounding the remains. Figure 6.24 displays the OCR data for the sampled soil column within the upper portions of the burial shaft.

The uppermost soil package provides dates for the fill and topsoil laid down during landscape activities of an existing nearby house. Interviews with long-term residents from this neighborhood confirmed that construction and landscaping of the property in question occurred in the late 1940s. The second soil package represents the *in situ* pedogenically altered soils infilling the burial shaft prior to the mid-20th-century landscaping. The maximum duration of these processes is 242YBP (AD 1708). This late date means that the individual buried here was likely still knowable through family oral histories. This opportunity to relate the present with the past and document cultural continuity is an important contribution to the local Native government and assisted in focusing their ethnohistorical search to identify the descendent families.

Unaltered sediments infilling the burial shaft below this soil package offer little information, and thus need not be further disturbed. Uniformity in the OCR ratio values, textures, manganese, and pH all suggest that these soils only have characteristics inherited from their original context prior to their initial excavation, mixing, and subsequently replacement into the shaft. The burial chamber rests below these unaltered fill sediments and did not need to be sampled or desecrated.

Examining Stratified Multicomponent Sites

Close-interval sampling along a vertical soil column helps define the related components of structural change in actively pedogenic soils as a system, establishing temporal contexts for archaeological artifacts and associated features. A comparison between samples in a soil column reveals certain individual processes and their relation to, and participation with, other pedogenic processes. Comparison of soil textures between samples shows evolving stone lines (coarse particles) and incipient argillic horizons (fine particles). Soil reactivity (pH), total organic carbon, and the OCR ratio describe consumption, digestion, and waste elimination processes.

Human occupation alters the trajectory of pedogenic processes in a soil through the introduction of additional organic matter (middens), pH change due to excessive acid or base loading, coarse particles (artifacts), and activities (both physical and chemical) that increase the weathering of clays from the surface or near-surface mineral soils through physical abrasion and removal of the protective vegetative boundary, and the deposition of artifacts. Immediately after deposition these materials and effects leave the cultural sphere and enter the pedological sphere, where soil formational processes further alter them, both chemically and physically. Such perturbational events as human activities remain evident in the soil profile as incipient

Soil Depth	pH	% Organic Carbon (LOI)	OCR Date	Very Coarse	Coarse	Medium	Fine	Very Fine	Coarse Silt	Fine Silt	Sample Id	% Oxidizable Carbon (WB)	OCR Ratio	Mn
4	6.4	6.332	3	10.455	7.934	12.770	32.382	12.993	9.386	14.081	4713	2.26	2.80	2.40
8	6.4	7.229	43	1.394	4.333	15.610	40.497	15.238	12.919	10.010	4714	2.93	2.47	2.14
12	6.4	5.614	141	8.927	8.076	15.118	32.198	14.121	13.801	7.759	4715	1.74	3.23	1.53
16	6.4	4.061	242	10.800	9.875	15.547	30.638	12.239	10.489	10.412	4716	1.27	3.20	0.99
20	6.3	2.609	226	8.109	10.154	17.743	34.932	11.487	9.105	8.470	4717	1.04	2.51	0.57
24	6.2	2.189	233	7.102	10.524	18.889	36.928	10.647	9.330	6.580	4718	0.80	2.74	0.61
28	6.0	1.498	231	8.899	12.032	21.467	38.087	8.665	5.673	5.178	4719	0.67	2.25	0.30
32	6.0	1.386	239	8.054	11.506	21.910	38.570	8.971	5.692	5.297	4720	0.54	2.57	0.33
36	6.0	1.784	240	7.925	11.240	22.317	36.541	9.249	6.345	6.383	4721	0.61	2.92	0.22
40	6.0	1.938	223	5.986	10.101	21.948	38.452	9.626	6.606	7.280	4722	0.99	1.96	0.26

Figure 6.24 OCR data for context around human burial.

evolving stone lines, argillic horizons, and uncharacteristic values for pH and total organic carbon compared to the normal expected trends.

Test pit 2 at archaeological site 16-VN-2920 in Fort Polk, Louisiana, demonstrates these dynamic soil processes as they relate to multiple occupations through time. Archaeological site 16-VN-2920 is a multicomponent site from which recovered cultural material, diagnostic stone tools and ceramics, suggests occupations during the Late Paleoindian (12,900 to 11,450 YBP), Late Archaic (5,700 to 3,200 YBP), and very late Late Woodland periods (after AD 1200). Artifact concentrations were focused around 10 cm, 40 cm, and 63 cm below the surface [318]. The OCR analyses identified three separate soil packages with coarse particle concentrations (evolving stone lines) at 10 cm, 40 cm, and 70 cm (Figure 6.25). Additionally, the OCR analysis suggests an aggraded and perturbated now-buried surface between 60 and 80 cm below the present surface.

Summary

Understanding soil complexity requires that soils be recognized as phenomena operating within a complex evolving system open relative to inputs of energy and matter from their environment. Furthermore, soils are not static but are constantly undergoing biochemical and physical changes according to their specific organizational design. The design is observable as complexes of biotic and abiotic self-organizing entities that relate at various levels and participate as components within each other.

Soil organizational design is defined not by the environment but by the soil itself. Again, it is not that Jenny's five factors of soil formation create soil, but rather that soil creates itself through its interactions as a self-organizing and self-maintaining system with its environment that includes these factors. Soils as self-organizing systems maintain themselves by dissipating energy into their environment, energy that is obtained from first-order perturbations. Additionally, emergent properties resulting from adaptive structural change in a soil are the result of interactions with second-order perturbations. This is evident in the unique values for the different variables from different soils and from different soil packages within the same soil. The general patterns of covariation between the variables identified for specific perturbations, however, support the notion that while the soil properties initiate possible responses to environmental perturbations, the specific environmental perturbations set limits on those options.

> "It seems that life in all its manifestations, from morphogenesis to symbolic thought, is governed by rules of the game which lend it order and stability but also allow for flexibility; and that these rules, whether innate or acquired, are represented in coded form on various levels of the hierarchy, from the genetic code to the structures in the nervous system associated with symbolic thought" [319; p. 43].

It would seem that such patterns are evident in abiotic self-organizing systems as well.

The various observed pedogenic processes explored in these case studies illustrate physical restructuring in terms of textural differentiation and chemical transformation

Soil Depth	pH	% Organic Carbon (LOI)	OCR Date	Very Coarse	Coarse	Medium	Fine	Very Fine	Coarse Silt	Fine Silt	Sample Id	% Oxidizable Carbon (WB)	OCR Ratio
10	3.4	2.215	1514	0.176	0.424	3.690	32.146	31.713	13.534	18.317	7154	0.83	2.67
20	3.3	0.960	2885	0.069	0.305	3.611	30.733	31.997	13.158	20.127	7155	0.32	3.00
30	3.1	0.630	4727	0	0.262	4.105	29.385	33.342	12.329	20.578	7156	0.20	3.15
40	2.9	0.530	6436	0.148	0.272	3.815	33.608	31.783	12.764	17.610	7157	0.20	2.65
50	2.7	0.430	8892	0.161	0.253	3.746	33.644	31.566	12.965	17.665	7158	0.13	3.31
60	2.6	0.504	9886	0.045	0.254	4.332	34.599	32.095	11.386	17.290	7159	0.07	7.20
70	2.5	0.407	9809	0.074	0.208	3.666	35.596	30.064	11.999	18.394	7160	0.07	5.81
80	2.4	0.473	9920	0.229	0.205	3.629	31.591	31.922	12.361	20.062	7161	0.07	6.76
89.5	2.3	0.364	12701	0.046	0.245	3.504	31.602	31.492	12.412	20.699	7212	0.12	3.17
99.5	2.3	0.352	16102	0.365	0.263	3.149	34.822	31.657	10.514	19.229	7213	0.10	3.52

Figure 6.25 OCR data for multicomponent site, 16-VN-2920.

in terms of organic carbon oxidation and associated manganese and pH values. Phrased differently, these observed pedogenic processes define a network of metabolic-like processes in which the function of each component is to participate in the production or transformation of other components in the soil network that together act to define and create soil. Soils worldwide can be treated as self-organizing systems operating within their own unique spatial and temporal scales and organized as systems definable by their structures, systems that can be understood and explained through processes that are comparable in their function and behavior to other recognized self-organizing systems such as living biological organisms.

7 Implications and Potentials of a Process-Oriented Soil Science

A distinction between open-systems and closed-systems views of soils and soil phenomenon has been made throughout this book. Soil science has become an integral part of many different fields of study, ranging beyond agronomy to include archaeology, physical and landscape geography, resource sustainability, and even climatology in terms of potential and long-term carbon sequestration. The importance and implications to these various fields of study in this distinction between understanding soils as open or closed systems will be briefly addressed in this chapter.

Modern soil science, as detailed in Chapter 2, is based on a closed-system paradigm: soils are the passive resultant of geological material weathered by the five soil-forming factors (climate, organism, relief, parent material, and time). The influence of these five factors results in a wide range of identifiable soils that can be described, classified, and inventoried. Such classifications allow for matching soil inventories to human uses. For example, a farmer might expect to harvest up to 24 tons per acre from an Agawam soil but only 16 tons per acre from a Deerfield soil. A Whitman soil is totally unsuited for growing corn due to the shallow depth of a perched water table [320].

Soils are critical components in research outside of agronomy and engineering, and modern soil science's epistemological limitations constrain the kinds of research questions these other disciplines are able to address. First off, as shown in Chapter 2, modern soil science evolved out of the study of and concern for plants, and not from a concern for the nature of soils themselves. The five-factor model and the chains of causality defined by the 13 paired processes of modern soil science evolved to simplify the complexity of one aspect of food production: soil as a plant growth medium. Second, modern soil science's classifications and inventories do not allow for the study of soils as dynamic self-organizing systems capable of changing or impacting on the five factors and the broader environmental context within which these interact (Chapter 3). Therefore, non-agronomic discipline needs for analyzing and understanding soils as a dynamic component within their research cannot be adequately addressed with the current models and paradigms of modern soil science. A process-oriented soil science, accepting of a dynamic universe of perturbations and sensitive to the behaviors of self-organizing systems, can inform research and policymaking in several significant areas of global concern. Figure 7.1 provides a small sampling of recent publications in some of these research areas.

Concepts	Geographic scale	References
Carbon sequestering	Local to global	(173)Amundson 2001; (141)Bailey, et al., 2002; (321)Bellamy, et al., 2005; (264)Bird, et al., 2003; (322)Capalbo, et al., 2004; (323)Carter 1996; (324)Christensen and Johnston 1997; (325)Copard, et al., 2006; (326)Corti, et al., 2002; (327)Davidson, et al., 2000; (328)Falloon and Smith 2003; (270)Fang, et al., 2005; (329)Foster 1981; (330)Gleixner, et al.,1999; (42)Janzen 2004; (331)Johnson, et al., 2005; (332)Kerner, et al., 2003; (333)Knorr, et al., 2005; (334,335)Lal 2004a & b; (336)Lal, R., et al., 1997; (337,338)Lal, et al., 2001a & b; (272)Mayorga, et al., 2005; (339)Oren, et al., 2001; (340)Pawlson 2005; (341,342)Petersen, et al., 2005a & b; (343)Richter, et al., 1995; (344)Rühlmann 1999; (345)Schimel, et al., 2000; (346)Schlesinger and Lichter 2001; (347)Simbahan and Dobermann 2006; (348)Simbahan, et al., 2006; (349)Sunda and Kieber 1994; (350)Theng, et al.,1992; (351)Torn, et al., 1997; (352)Townsend, et al., 1995; (353)Treseder and Allen 2000; (354)Trumbore 1997; (355)Vagen, et al., 2005; (356)Wiseman and Puttmann 2006; (357)Zhou, et al., 2006.
Soil health indices	Local to global	(358)Barr 2003; (359)Barriosa, et al., 2006; (360)Doran 2002; (361)Doran and Parkin 1994; (362)Doran, et al., 1994; (363)Doran, et al., 2002; (364)Haberern 1992; (365)Hund-Rinke, et al., 2003; (366)Lal 2004; (336)Lal, et al., 1997; (367)Larson and Pierce 1994; (368)Melitz 1986; (369)O'Connell, et al., 2000; (370)Pankhurst 1994; (371)Pierce, et al., 1983; (372)Qi, et al., 1994; (373)Sanchez, et al., 2003; (374)Seryi 1996; (375)Stewart 1968; (376)USDA Natural Resources Conservation Service 1996; (377)Wall and Virginia 1999.
Waste and pollution management	Local to regional	(378)Entry, et al., 1998; (379)Fresquez, et al., 1986; (380)Garcia, et al., 1998; (278)Jones and Muehlchen 1994; (381)Khan, et al., 1998; (382)Melgar, et al., 1998; (383)Moller, et al., 2005; (384)Munro, et al., 1997; (385)Murray, et al., 1997; (386)Nofziger, et al., 1996; (387)Rikken and Van Rijn 1993; (388)Rundel and Kogel-Knabner 2004; (389)Schmidt 1998; (390)Schmidt, et al., 2000; (391)Sipos, et al., 2005; (392)Sodergren 1968; (393)Steenhuis, et al., 1998; (394)Stolpe, et al., 1998; (395) Tsay, et al., 1970; (396)Vaishampayan and Hemantaranjan 1984.
Agriculture practices	Local to global	(397)Ahuja 2003; (398)Bakker 1979; (399)Beavis and Mott 1996; (198)Blum, et al., 2002; (400)Capriel 1997; (401)Cools, et al., 2003; (402)Flores, et al., 2005; (403)Fründ, et al., 1994; (404)Janzen, et al., 1997; (405)Körschens 1998; (406)Lal 1999; (407)Leifeld and Kogel-Knabner 2005; (408,409)Otterman 1974; 1977; (410)Paré, et al., 1999; (411)Rogers and Burns 1994; (412)Sheley, et al., 1996; (413)Six, et al., 2000; (414)Storie 1933; (415)Storie 1978; (416,417)Tongway 1994; 1995; (418,419)van Lanen, et al., 1992a & b; (420)von Lützow, et al., 2001; (421)Wu, et al., 2003; (422)Yemefack, et al., 2005; (423)Young 1989.
Designing sustainable geo-ecosystems	Regional	(424)Blum and Santelises 1994; (425)Bouma 2001; (426)Brown, et al., 2000; (427)Greenland and Szabolics 1994; (428)Körschens, et al., 1998; (429)Land and Water Development Division 1998; (430)Lemieux 1997; (431)Mitchell, et al., 1998; (432,433)Oldeman 1988; 1994; (434)Oldeman, et al., 1991; (435)Richter and Markewitz 2001; (436)Smith, et al., 1992; (437)Soil Resources Inventory Group 1981; (438)Steiner 1996; (439)Thwaites and Slater 2000; (440)van Diepen, et al., 1991; (441)Wilson and Bryant 1997; (442,443)Zhu 1997; 2000; (444)Zinck 1988.
Archaeology extra terrestrial life	Local galactic	(299)Hughes and Lampert 1977; (251)Saunders 1997; (445)Schiffer 1987; (206)Blackwell 2000; (117)Ferris, et al., 1996; (446)Gray and Shear 1992; (185)Joyce 2002; (447)Markewitz 1997; (448)Mojzsis, et al., 1996; (449)Navarro-Gonzales, et al., 2003; (450)Poulet, et al., 2005; (451)Stixrude and Peacor 2002.

Figure 7.1 Citations sorted according to research topics and the geographical scale at which they are expressed.

Carbon Sequestration and Climate Change

Much of the research into the use of soil for carbon sequestering in a response to global warming has been constrained by the closed-system approach of modern soil science to discussions on, for example, the suitability of forest versus agricultural soils for carbon sequestration, and quantitative measures of the total carbon content/capacity of each. The latter ignores the fact that labile carbon molecules will quickly oxidize and return to the atmosphere. As presented in Chapter 4, soil organic carbon takes on many molecular forms, ranging from labile raw organic matter through inert lignite (brown coal). For effective long-term soil carbon sequestration studies, an understanding of a soil's ability to create and bind recalcitrant and inert carbon molecules is required.

Research into the use of soil for carbon sequestering in a response to global warming can be informed at global as well as regional and local scales by understanding the metabolic pathways by which soil mineralizes (returning carbon dioxide to the atmosphere), alters to biologically resistant and inert forms, and sequesters carbon within multiple components and strata. This understanding is essential to effective use of soil as a major carbon sink. As shown in Chapter 4, Vertisols and Spodisols are poorly adapted to the task of effective carbon metabolization, leading to long-term carbon sequestration. Molisols and Alfisols, with their high base status and montmorillonite clays, however, are capable of quickly transforming labile carbon molecules into more resistant and inert forms. Research on engineering better carbon-sequestering soils through improved metabolic pathways becomes one of many new areas of study.

Soil Health Indices

Increased desertification, resulting from climate change and agriculturally marginal land degradation, and increased urbanization and sprawl continue to limit land available for agriculture. New lands capable of offsetting this ongoing loss are simply no longer available on this planet. Recognition of these facts has led to the development of soil health indices to track and monitor changing soil health in the increasingly scarce agricultural and forest soils. At present, soil health indices simply inventory and quantify key constituents, such as available phosphorus and degree of microbial biodiversity, or measure carbon dioxide emissions as a measure of microbial activity. Thus the very idea of soil health is based on its potential productivity and not on how the soil itself is organized.

Soils at various taxonomic levels display unique carbon quality and quantity profiles that can be used to define soil health in terms of the soil's self-organization. Soil health indices that assess conditions at the local scale and inform inventories at regional and global scales can be enhanced by a better understanding of whether a soil is actually metabolizing efficiently within its specific environmental context.

Waste and Pollution Management

Increased industrialization coupled with urbanization has led to increased wastes and pollution degrading water and land. Brownfields within cities, toxic lands resulting from overreliance on petrochemical fertilization that accumulate at the margins of many agricultural areas, landfills with leachate plumes, and many other point and non-point pollution problems have a negative impact on soil health and have become common land use planning problems. Waste and pollution management are concerns operating at local scales where soil series data provide detailed inventories and use classifications, none of which indicate suitability for waste and pollution. As the quantity and distribution of wastes and pollutants continue to increase, this becomes a management concern at regional scales as well. Soil classifications and inventories are of little use at these scales without an understanding of how specific soils might relate and interact, mobilizing certain chemicals through waste-elimination physiological process, or fixing by way of transferring the other chemical compounds to function as a new constituent part of the soil system.

Understanding how soil concentrates, removes, and stabilizes human pollutants in general and specific pollutants in particular can lead to management policies better able to rank and triage mitigating actions. Assessments tied to the soil's health index based on its self-organizing characteristics can point out deficiencies preventing or limiting a soil's ability to effect control of pollutants. Conversely, such assessments can direct actions to enhance a soil's capacity to mitigate the effects of pollutants.

Wastes and pollutants are perturbations to the soil system (Chapter 3). As such, it is important to understand how the soil as a system will respond, both in terms of how it will be affected and how it will affect the perturbation and surrounding environment. Current use of soil science in waste and pollution management is generally restricted to the experiment proposed in Chapter 3 designed to determine if perturbation "A" results in soil "B," where the scientist subjects a soil to the perturbation and waits to observe its effect on the soil's development, without considering the multitude of other first- and second-order perturbations that might apply to this open system.

Soil Geography

Jenny [30] makes the distinction between soil geographers, those who map the locations, variability, and relationships between soil across the landscape, and soil functionalists, the soil scientists who identify, conduct the laboratory analyses, classify soil series, and determine their use suitability. Together, these two groups compose the functional modern soil science. However, the kinds of research questions available to the soil geographer depend on the classifications and inventories defined by the soil functionalist.

Currently soil landscape assessments are generally limited to landform studies based on catenas, a series of related soil types that differ only in terms of one of the five factors of soil formation. For example, a topo-drainage catena where slope and its effect on depth to water table explain the different soils that evolved from a common initial event, or a topo–chrono sequence, such as pediments evolving through varying periods of time. Alternatively, common genesis of deposition, such as glacial, lacustrine, or riverine landscape, is used to describe collections of landforms. As such, studies have been limited to local and regional scales and related landscape component variability, such as biotic complexity, are similarly restricted [452].

In Chapter 6, I show how recurrent perturbations evident in different landforms (dunes, riverbanks, and lagoon edges) across landscapes are structurally coupled with their environment, and that analyses viewing them as open dynamic systems can explain common changes such as climatic variability across broadly defined regions. Additionally, I demonstrate how specific landforms can be distinguished at global scales (dunes and levees in North and South America) based on the characterization of their total organic carbon and percent oxidizable content, and the quality of that carbon as expressed by the OCR ratio, and that these data can be used to deduce differing initial conditions and genesis of these landforms.

Designing Sustainable Geo-ecosystems

Ecological studies tend to treat soil as the environmental context within which biological processes are enacted, and most research questions focus on factors affecting biodiversity on land, in the waters and air, and in the soil [453,454] with little regard to the soil itself. Large-scale environmental diversity has been shown to require a constant low level of perturbations [68,69]. These perturbations within ecosystems, however, can also alter the nature and properties of soil [455]. Invasive vegetation has been shown to affect the soil-food-web [456], as well as increasing or diminishing soil nitrogen and carbon balances and pH values [457,458]. Additionally, soil structure and plant available nitrogen from organic sources are affected by mycorrhizal fungi, whose presence or absence in an ecosystem is dependent on soil disturbance levels [459,460]. These ecological studies stress the need for better understanding of the complex feedback behavior between plant and soil systems.

Designing sustainable geo-ecosystems across regional landscapes can benefit from an integrative and holistic approach that treats all soils appropriate to their effective metabolic processes and not just their capacity for production. Considering self-organizing behavior at various scales simultaneously can result in both healthy sustainable soils and healthy sustainable landscapes.

Agricultural Practices

Agricultural practices focused exclusively on crop yields at a local scale can be transformed to practices focused on sustainable production at a global scale through a better understanding of what soils need beyond specific constituent amendments to effectively carry out their metabolic processes. The emergent field of agroecology, treating soil as a living ecological system rather than a factory in an industrial system, offers new process-oriented perspectives on food production [461]. Programs sensitive to the complex relationships existing between soil components are already leading to the inoculation of forests with mycorrhizal fungi (mycoforestry), thereby enhancing tree health and retarding soil erosion potential. Organic gardeners have long recognized the importance of mycorrhizal fungi for pest control, as well as for enhancing crop growth [86,462,196].

Geo-archaeology

Geo-archaeology in general provides archaeological research with information on landform genesis at the local scale, and landscapes at more regional scales. These data have informed research questions on site selection behavior, resource exploitation strategies, and cultural landscapes. Recently, soil micro-morphology has reveled new contexts and artifact types at the micro scale.

Throughout this book, I have shown how the archaeological context of artifacts can be better understood by considering soils' self-organizing behavior. As demonstrated in Chapter 3, the punctuated dynamics of soil formation resulting from second-order perturbations may be thought of as resulting in separate packages of soil deposits, each expressing unique and related spatial and temporal contexts. Beyond site formation, a process-oriented soil science expands archaeological research into the post-occupation transformations of the archaeological record [303].

Pollen, gastropods, and phytoliths used in paleoenvironmental studies have similar soil contextual issues. These environmentally sensitive indicators cannot be assumed to have a primary context as they may be translocated through the soil body at various rates, depending on their size, as a result of soil textural sorting, waste elimination, and other metabolic processes.

The dynamic effects of soil perturbations are echoed in various archaeological studies. For example, Aims [463] recommends that increased attention be paid to production processes themselves in the archaeological debate on subsistence strategies between pure hunting and gathering and agriculture-focused peoples. Such processes are most likely to be found as alterations or perturbations evident in the soil; for example, the anthropogenic development of black earths [79,464], engineering soil expansion to increase a critical resource niche [465], or an evident change in a soil's structural characteristics resulting from deforestation during the Early Neolithic period in central Europe [304]. Residual soil fertility can also be used for reconstructing past land use in paleo-ecological studies [466].

Astrobiology

A process-oriented soil science can inform astrobiology and the search for extraterrestrial life by looking for evidence of complexity and self-organization not just limited to organic molecules but also in soils and the processes exhibited by the planet itself—for example, comparing the millions of soil series found on Earth, a planet infused with biological life, to the two soils of Mars, where despite much effort no evidence of life has been found. Self-organizational behavior, it would appear, is an emergent property that is reflected across all scales from the extremely small to the extremely large, and its emergence is dependent on those initial conditions that set constraints on how systems can behave.

Because initial conditions will constrain the organization of systems from a global scale, it should not be expected that life on planet X in galaxy Y will be composed of the same constituent materials and organized in the same way as it does here on Earth.

Soil development on the Moon, Mars, and Venus has already been shown to follow totally different trajectories than those evident in even the earliest dated paleosols on Earth. Precambrian paleosols are similar in many ways to deeply weathered soils at the land surface today. Nothing in the Precambrian rock record has yet been found comparable with the soils rich in glass and agglutinates on the Moon or the glazed and metamorphosed soils thought to be forming on Venus. Known Precambrian paleosols are neither as oxidized and sulfate-rich as the soils of Mars, nor as different from modern soils as they could be [1].

These dispersed and varied areas of scientific study share the concern that the paradigm and models of traditional soil science lack the ability to adequately address issues of soil dynamics, environmental integration, and change. Unexplainable research results obtained from traditional soil studies applied to nontraditional soil phenomena in physical geography, archaeology, and ecology speak to the current need for soil science to move beyond description and classification and into a dynamic process-oriented soil science capable of providing explanations.

8 Summary

The preceding chapters have examined the history and development of modern soil science, and the discovery, development, and some of the applications in recent archaeological, geomorphologic, and pedological studies of the OCR procedure. The OCR formula was initially developed to model unexpected and unexplained soil characteristics revealed in the results from common soil analyses. In the search for explanations, it became necessary to question many assumptions made about soils and how modern soil science treats soils. First, soil's unique spatial and historical context needed to be factored into all interpretations. Next, soil's dynamic context in a universe of perturbations required reconsideration of how soils are conceived of and described. Soils do not behave as static inert geological detritus affected by climate, organisms, relief, and parent material through time, but instead soils behave as any self-organizing system, dynamically interrelating with their environment. Recognition of this dynamic behavior required a reexamination of how science in general thinks and how modern soil science specifically evolved its basic paradigms and models. Through this reexamination, it became necessary to explore and develop a process-oriented soil science that could be built onto traditional soil science, but built in such a way as to incorporate the inherent physiological dynamics evidenced by soils as self-regulating, self-maintaining, and self-replicating systems.

What is wrong with modern soil science? In and of itself, there is nothing wrong with it. Its contributions to agriculture and engineering over the past century have been invaluable, allowing the human species to grow exponentially and to expand into formerly uninhabitable regions of this planet. But understanding soils is a critical component of research outside of agronomy, and modern soil science's inability to provide explanations describing soil behavior is limiting these studies.

It is not that soils themselves are different within the different disciplines of study. Rather, it is the nature of the questions being asked and the contextual perspectives applied to these questions that expand the view of soils beyond that historically required by soil science. These different perspectives need not be seen as divergent competing views on the nature of soil. Rather, they can be convergent to a more complete and integrated view of soil as it exists within a holistic reality undivided by disciplinary concerns.

Convergence, however, is not the current condition between these various disciplines. The present divergence between modern soil science as a descriptive and classificatory discipline, and the need for explanations in physical geography, archaeology, and ecology varies according to the degree to which each of these disciplines consider dynamic system-oriented processes as an essential epistemological

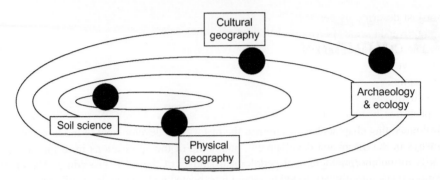

Figure 8.1 Schematic representation of decreasing utility of the five-factor model and universal definition of soil in studies by disciplines relative to soil science.

component (Figure 8.1). Examples of this divergence can be found within the recent literature of these disciplines (see Figure 7.1).

In physical geography, Phillips [467,468] attempts to integrate deterministic chaos with the five-factor model to explain coastal plain soil variability within single landforms. The ontology and epistemology of the complexity of landscapes in process-oriented physical geography [469] are considered to be more basic than substantive material objects [470]. However, Huggett [471] argues in response that while Jenny's [472] recasting of his original 1941 model may appear to consider the importance of initial conditions and historical contingency inherent in open-system dynamics, "Jenny's revised model of pedogenesis remained firmly position in the classical thermodynamic tradition" [471; p. 24].

Johnson and coworkers [33,37,473–475] see the traditional distinction between soil horizon formation and destruction by perturbations as a false dichotomy obscuring the more typical dynamics of soil formation. Their theory of dynamic denudation posits that in the course of normal biochemical and biomechanical processes in the soil, coarse particles such as broken sapolitic (rotten) bedrock and cultural artifacts are translocated downwards through the soil profile to form stone line features [37].

In fact, as demonstrated in this book, various spatially separated components are temporally separated as well, each drifting downwards (or upwards as is the case for desert pavements and carbonates) through the soil profile from a common surface. Thus, the soil profile has to be evaluated as an ongoing continuum of structural change within definable domains, or packages. Soil cannot be understood solely by a description of its constituent parts, but must be holistically viewed in terms of the spatial and temporal relationships that exist between various parts; in other words, as a dynamic open system [108].

A process-oriented soil science, accepting of a dynamic universe of perturbations and sensitive to the behaviors of self-organizing systems, can inform current research and policymaking in several significant areas of global concern.

What began a quarter-century ago as a search for an explanation of the unexpected results obtained from simple soil analyses culminates here as the first steps in the development of a new way to think of soil and its characterization, to explain and not just describe its geographical distribution across space at local, regional, global, and universal scales [476].

The OCR formulation provides soil science with a first enumeration, a modeling tool if you will, for the necessary mapping out of appropriate contexts for these various samplings. However, as a modeling tool, it is important to consider that it too is a reduction of reality, and needs to be challenged, augmented, and improved in ways appropriate to including more of nature's complexity. Models, theories, and epistemologies inherently constrain the kinds of observations that are possible, and should always be reviewed and critiqued as a part of ongoing scientific inquiry. Use of the OCR formulation does not allow us to undertake soil studies with the certainty that we are on the edge of some brave new world, mystically armed with a new technique that will answer all of our questions. Rather, as I hope I have demonstrated throughout this work, we undertake soil studies, and all of science, with doubt and the inherent vulnerability of confusion that we face as paradigms are questioned. It is only from this position of doubt and our human condition of confusion that we can make new beginnings in our exploration of nature, unfettered by constraining beliefs.

Bibliography

[1] Retallack GJ. Soils of the Past: An Introduction to Paleopedology. Oxford, UK: Blackwell Science; 2001.
[2] Ball DF. Loss-on-ignition as an estimate of organic matter and organic carbon in non-calcareous soils. J. Soil Sci. 1964;15(1):84–92.
[3] Walkley A. An examination of methods for determining organic carbon and nitrogen in soils. J. Agr. Sci. 1935;25:598–609.
[4] Walkley A, Black IA. An examination of the Degtjareff method for determining soil organic matter and a proposed modification of the chromic acid titration method. Soil Sci. 1934;37:29–38.
[5] Amadon JF. Soil organic matter characterization and nitrogen mineralization. Masters Thesis, Department of Plant and Soil Sciences. Burlington, VT: University of Vermont; 1979.
[6] Bonemisza E, Costenala M, Alvarado A, Ortega EJ, Vasquez AJ. Organic determination by the Walkley-Black and dry combustion methods in surface soils and depth profiles from Costa Rica. Soil Sci. Soc. Am. Proc. 1979;26:254.
[7] Walkley A. A critical examination of a rapid method for determining organic carbon in soils – effect of variations in digestion conditions and of inorganic soil constituents. Soil Sci. Soc. Am. J. 1947;63:251–64.
[8] Baveye P. Comment on "Modeling soil variation: past, present and future" by GBM. Heuvelink and R. Webster. Geoderma 2002;109(3–4):289–93.
[9] Bourennane H, Salvador-Blanes S, Cornu S, King D. Scale of spatial dependence between chemical properties of topsoil and subsoil over a geologically contrasted area (Massif central, France). Geoderma 2003;112(3–4):235–51.
[10] Casey ES. Between geography and philosophy: what does it mean to be in the place-world? Ann. Assoc. Am. Geogr. 2001;91(4):683–93.
[11] Wagenet RJ. Bouma J, editors. Geoderma, 60; 1993 p. 89–107.
[12] Webster R, McBratney AB. Soil segment overlap in character space and its implication. J. Soil Sci. 1981;32:133–47.
[13] Veblen T. The beginnings of ownership. Am. J. Sociol. 1898;4:352–65.
[14] Jarrett B. FeudalismKnight K, editor. The Catholic Encyclopedia, vol. VI. New York: Robert Appleton Company; 1909. p. 15. [accessed 01.08.10]. http://www.newadvent.org/cathen/06058c.htm
[15] Briffa KR, Bartholin TS, Eckstein D, Jones PD, Karlen W, Schweingruber FH, Jetterberg P. A 1,400-year tree-ring record of summer temperatures in Fennoscandia. Nature 1990;346:434–9.
[16] Keigwin LD. The Little Ice Age and Medieval Warm Period in the Sargasso Sea. Science 1996;274:1503–8.
[17] Stuiver M, Reimer PJ, Bard E, Beck JW, Burr GS, Hughen KA, Kromer B, McCormac FG, Plicht J vd, Spurk M. IntCal98 radiocarbon age calibration, 24,000–0 cal BP. Radiocarbon 1998;40(3):1041–83.
[18] Broecker WS. Was a change in thermohaline circulation responsible for the Little Ice Age? Proc. Natl. Acad. Sci. U. S. A. 2000;97(4):1339–42.

[19] Fagan B. The Little Ice Age: How Climate Made History 1300–1850. New York: Basic Books; 2000.
[20] Bamforth DB, Bleed P. Technology, flaked stone technology, and risk. Barton CM, Clark GA, editors. Rediscovering Darwin: Evolutionary Theory in Archaeological Explanation, vol. 7; 1997 p. 109–40.
[21] Dear P. Revolutionizing the Sciences: European Knowledge and Its Ambitions 1500–1700. Princeton, NJ: Princeton University Press; 2001.
[22] Bacon F. Novum Organum. Chicago, IL: Open Court; 1994 [1620]. Translated and edited by Peter Urbach and John Gibson
[23] Brinton C. EnlightenmentEdwards P, editor. The Encyclopedia of Philosophy, vol. 2. New York: Macmillan Publishing Co. The Free Press; 1967. p. 519–25.
[24] Locke J. Second Treatise of Civil Government. New York: MacMillian Press; 1986 [1690].
[25] Tuttle H. History of Prussia to the Accession of Frederic the Great: 1134–1740. Boston, MA: Houghton, Mifflin and Co; 1884.
[26] Steven S. Larding the Lean Earth: Soil and Society in Nineteenth-Century America. New York: Hill and Wang: A division of Farrar, Straus and Giroux; 2002.
[27] Glinka KD. Dokuchaiev's ideas in the development of pedology and the cognate sciences Russian Pedological Investigations No. 1. Leningrad: USSR Academy of Sciences; 1927.
[28] Dokuchaev VV. (1967 [1898]). Russian Chernozem (Russkii Chernozem). Collected Writings (Sochineniya), vol. 3. Springfield, VA: U.S. Department of Commerce.
[29] Basinski JJ. The Russian approach to soil classification and its recent development. J. Soil Sci. 1959;10:14–26.
[30] Jenny H. Factors of Soil Formation – A System of Quantitative Pedology. New York: McGraw-Hill; 1941.
[31] Buol SW, Hole FD, McCracken RJ, Southard RJ. Soil Genesis and Classification. Ames, IA: The Iowa State University Press; 1997.
[32] Hole FD. A classification of pedoturbations and some other processes and factors of soil formation in relation to isotropism and anisotropism. Soil Sci. Soc. Am. J. 1961;91:375–7.
[33] Johnson DL, Watson-Stegner D. Evolution model of pedogenesis. Soil Sci. Soc. Am. J. 1987;143:349–66.
[34] Wood WR, Johnson D. A survey of disturbance processes in archaeological site formationSchiffer M, editor. Advances in Archaeological Method and Theory, vol. 1; 1978. p. 315–81.
[35] Brown LR. Seeds of Change: The Green Revolution and Development in the 1970's. New York: Praeger Publishers; 1970.
[36] Shaw DJB, Oldfield JD. Landscape science: a Russian geographical tradition. Ann. Assoc. Am. Geogr. 2007;97(1):111–26.
[37] Johnson DL. Biomechanical processes and the Gaia paradigm in a unified pedo-geomorphic and pedo-archaeological framework: dynamic denudation. In: Foss J, Timpson M, Morris M, editors. First International Conference on Pedo-Archaeology. Knoxville, TN: University of Tennessee; 1992. p. 41–67.
[38] Taleb NN. The Black Swan: The Impact of the Highly Improbable. New York: Random House; 2007.
[39] Tompkins P, Bird C. Secrets of the Soil. Anchorage, AL: Earthpulse Press; 1998.
[40] Pollan M. In Defense of Food: An Eater's Manifesto. New York: Penguin Press; 2008.
[41] Kurlansky M. Salt: A World History. New York: Penguin Books; 2002.

[42] Janzen HH. Carbon cycling in earth systems – a soil science perspective. Agr. Ecosyst. Environ. 2004;104:399–417.
[43] Descartes R. (1984 [1644]). Principles of Philosophy. Boston, MA: D. Reidel.
[44] Willstatter R. Untersuchungen uber Enzyme. Berlin: Springer; 1928.
[45] Midgley M. The Myths We Live By. New York: Routledge; 2003.
[46] Capek M. ChangeEdwards P, editor. The Encyclopedia of Philosophy, vol. 2. New York: Macmillan Publishing Co. & The Free Press; 1967. p. 75–9.
[47] Leibniz GW. (1991 [1646–1716]). Discourse on Metaphysics & Other Essays. Indianapolis, IN: Hackett Publishing Company, Inc; 1991. p. 1646–1716.
[48] Needham J. Science and Civilisation in China. Cambridge: Cambridge University Press; 1969.
[49] Wilhelm R. The I Ching or Book of Changes. Princeton, NJ: Princeton University Press; 1950.
[50] Emmet DM. Alfred North Whitehead. In: Edwards P, editor. The Encyclopedia of Philosophy, vol. 8. New York: Macmillan Publishing Co. & The Free Press; 1967. p. 290–7.
[51] Whitehead AN. Process and Reality. New York: The Free Press; 1978.
[52] Dewey J. (1997 [1910]). How We Think. Mineola, NY: Dover Publications.
[53] Wiener PP, editor. Charles S. Peirce: Selected Writings (Values in a Universe of Chance). New York: Dover Publications; 1958.
[54] Foucault M. The Archaeology of Knowledge & the Discourse on Language. New York: Pantheon Books; 1972.
[55] Midgley M. Science and Poetry. London: Routledge; 2001.
[56] Rorty R. Philosophy and the Mirror of Nature. Princeton, NJ: Princeton University Press; 1979.
[57] Buol SW, Hole FD, McCracken RJ. Soil Genesis and Classification. Ames, IA: Iowa State University Press; 1980.
[58] Maturana H, Varela F. The Tree of Knowledge: The Biological Roots of Human Understanding. Boston, MA: Shambhala; 1998.
[59] Meentemeyer V. Geographical perspectives of space, time, and scale. Landscape Ecol 1989;3(3/4):163–73.
[60] Phillips JD. Entropy analysis of multiple scale causality and qualitative causal shifts in spatial systems. Prof. Geogr. 2005;57(1):83–93.
[61] Southworth J, Cumming G, Marsik M, Binford M. Linking spatial and temporal variation at multiple scales in a heterogeneous landscape. Prof. Geogr. 2006;58(4):406–20.
[62] Young K, Aspinall R. Kaleidoscoping landscapes, shifting perspectives. Prof. Geogr. 2006;58(4):436–47.
[63] Finke PA, Bouma J, Stein A. Measuring field variability of disturbed soils for simulation purposes. Soil Sci. Soc. Am. J. 1992;56(1):187–92.
[64] Pluis JLA, De Winder B. Natural stabilization. Catena Supplement. 1990;3:195–208.
[65] Rapport DJ, Whitford WG. How ecosystems respond to stress. Common properties of arid and aquatic systems. BioScience 1999;49(3):193–203.
[66] Maturana H, Varela F. Autopoiesis and Cognition: The Realization of the Living. Boston, MA: D. Reidel; 1980.
[67] Sagan D, Margulis L. Gaia and philosophy. In: Margulis L, Sagan D, editors. Dazzle Gradually: Reflections on the Nature of Nature. White River Junction, VT: Chelsea Green Publishing; 2007. p. 172–84.
[68] Robbins P, McSweeney K, Waite T, Rice J. Even conservation rules are made to be broken: implications for biodiversity. Environ. Manage. 2006;37(2):162–9.

[69] Wardle DA, Walker LR, Bardgett RD. Ecosystem properties and forest decline in contrasting long-term chronosequences. Science 2004;305:509–13.
[70] Sparling G, Ross D, Trustrum N, Grnold A, West W, Speir T, Schipper L. Recovery of topsoil characteristics after landslip erosion in dry hill country of New Zealand, and a test of the space-for-time hypothesis. Soil Biol. Biochem. 2003;35:1575–86.
[71] Bouma J, Droogers P. Comparing different methods for estimating the soil moisture supply capacity of a soil series subjected to different types of management. Geoderma 1999;92(3–4):185–97.
[72] Berg AW. Formation of Mima mounds: a seismic hypothesis. Geology 1990;18:281–4.
[73] Lyford WH, MacLean DW. Mound and pit microrelief in relation to soil disturbances and tree distribution in New Brunswick, Canada. Harvard Forest Paper 1966;15:1–18.
[74] Allgood FP, Gray F. Genesis, morphology, and classification of mounded soils of eastern Oklahoma. Soil Sci. Soc. Am. Proc. 1973;37(5):746–53.
[75] Allgood FP, Gray F. An ecological interpretation for the small mounds in landscapes of eastern Oklahoma. J. Environ. Qual. 1974;3(1):37–41.
[76] Cox GW. Soil mining by pocket gophers along topographic gradients in a Mima mound field. Ecology 1990;71:837–43.
[77] Cox GW, Berg AW. Comment and reply on "Formation of Mima mounds: a seismic hypothesis". Geology 1990;18(12):1259–61.
[78] Johnson DL, Horwath J, Johnson DN. Mima and other animal mounds as point-centered biomantles. *Geol. Soc. Am. Abstr. Programs* 2003;35(6):258.
[79] Glaser B, Balashov E, et al. Black carbon in density fractions of anthropogenic soils of the Brazilian Amazon region. Org. Geochem. 2000;31:669–78.
[80] Kristiansen SM. Present-day soil distribution explained by prehistoric land-use: Podzol-Arenosol variation in an ancient woodland in Denmark. Geoderma 2001;103(3–4):273–89.
[81] Shubert LE, Starks TL. Diagnostic aspects of algal ecology in disturbed lands. In: Tate RL, Klein DA, editors. Soil Reclamation Processes. Microbiological Analyses and Applications. New York: Marcel Dekker, Inc.; 1985. p. 83–106.
[82] Starks TL, Shubert LE. Algal colonization on a reclaimed surface-mined area in western North DakotaWali MK, editor. Ecology and Coal Resource Development, vol. 2. New York: Pergamon Press; 1979. p. 652–60.
[83] Starks TL, Shubert LE. Colonization and succession of algae and soil-algal interactions associated with disturbed areas. J. Phycol. 1982;18:99–107.
[84] Bridges M, De Bakker H. Soil as an artefact: human impacts on the soil resource. Land (Chatham) 1997;1(3):197–215.
[85] Dudal R, Nachtergaele F, Purnell M. The human factor of soil formation. Symposium 18, Vol. II, paper 93. Transactions 17th World Congress of Soil Science, Bangkok; 2002.
[86] Faulkner E. Plowman's Folly. Norman, OK: University of Oklahoma Press; 1974.
[87] Kade A, Warren SD. Soil and plant recovery after historic military disturbances in the Sonoran Desert, USA. Arid Land Res. Manag. 2002;16:231–43.
[88] Paz-González A, Vieira SR, Taboada Castro MT. The effect of cultivation on the spatial variability of selected properties of an umbric horizon. Geoderma 2000;97(3–4):273–92.
[89] Preston CM, Newman RH, Rother P. Using 13C CPMAS NMR to assess effects of cultivation on the organic matter of particle size fractions in a grassland soil. Soil Sci. 1994;157:26–35.
[90] Czimczik CI, Preston CM, Schmidt AWI, Werner RA, Schulze E. Effects of charring on mass, organic carbon, and stable carbon isotope composition of wood. Org. Geochem. 2002;33:1207–23.

[91] Golchin A, Clarke P, Baldock JA, Higashi T, Skjemstad JO, Oades JM. The effects of vegetation and burning on the chemical composition of soil organic matter in a volcanic ash soil as shown by 13C NMR spectroscopy. I. Whole soil and humic acid fraction. Geoderma 1997;76:155–74.

[92] Kershaw KA, Rouse WR. The impact of fire on forest and tundra ecosystems final report 1975. Ottawa, ON: Department of Indian and Northern Affairs; 1976.

[93] Kinnell PIA, Chartres CJ, Watson CL. The effects of fire on the soil in a degraded semi-arid woodland, II. Susceptibility of the soil to erosion by shallow rain-impacted flow. Aust. J. Soil Res. 1990;28:779–94.

[94] Pyne SJ. Fire in America. Princeton, NJ: Princeton University Press; 1988.

[95] Runkle JR. Disturbance regimes in temperate forests. In: Pickett STA, Whitep PS, editors. The Ecology of Natural Disturbance and Patch Dynamics. Orlando, FL: Academic Press; 1985. p. 17–33.

[96] Bird M, Kracht O, Derrien D, Zhou Y. The effect of soil texture and roots on the stable carbon isotope composition of soil organic carbon. Aust. J. Soil Res. 2003;41:77–94.

[97] Sencindiver JC, Ammons JT. Minesoil genesis and classification. In: Barnhisel RI, Daniels WL, Darmondy RG, editors. Reclamation of Drastically Disturbed Lands. Madison, WI: American Society of Agronomy; 2000. p. 595–613.

[98] Frink DS. Demonstrating emergent properties in dynamic systems. *J. Coll. Sci. Teach.* XXXIX 2009;1:68–70.

[99] Macklem PT. Emergent phenomena and the secrets of life. J. Appl. Physiol. 2008;104:1844–6.

[100] Ruellan A. The history of soils: some problems of definition and interpretation. In: Yaalon DH, editor. Paleopedology: Origin, Nature and Dating of Paleosols. Jerusalem: Israel University Press; 1971. p. 3–13.

[101] Yaalon DH. Soil-forming processes in time and space. In: Yaalon DH, editor. Paleopedology: Origin, Nature and Dating of Paleosols. Jerusalem: Israel University Press; 1971. p. 29–39.

[102] Phillips JD. Simplexity and the reinvention of equifinality. Geogr. Anal. 1997;29(1):1–15.

[103] Loranz E. Deterministic nonperiodic flows. J. Atm. Sci. 1963;10(2):130–44.

[104] Sagan D, Schneider ED. The pleasures of change. In: Margulis L, Sagan D, editors. Dazzle Gradually. White River Junction, VT: Chelsea Green Publishing; 2007. p. 223–37.

[105] Prigogine I, Stengers I. Order Out of Chaos: Man's New Dialogue with Nature. New York: Bantam Books; 1984.

[106] Davies P. The Cosmic Blueprint: Order and Complexity at the Edge of Chaos. London: Penguin; 1989.

[107] Albrecht A. Cosmic inflation and the arrow of time. In: Barrow JD, Davies PCW, Harper CL, editors. Science and Ultimate Reality: From Quantum to Cosmos. Cambridge: Cambridge University Press; 2003. p. 363–401.

[108] Laszlo E. Introduction to Systems Philosophy: Toward a New Paradigm of Contemporary Thought. New York: Harper & Row; 1972.

[109] McAuliffe JR. Desert soils. In: Phillips SJ, Comus PW, editors. A Natural History of the Sonoran Desert. Tucson, AZ: Arizona-Sonora Desert Museum Press; 2000. p. 87–104.

[110] Jantsch E. The Self-Organizing Universe. New York: Pergamon; 1980.

[111] Margulis L, Sagan D. What is Life? Berkeley, CA: University of California Press; 1995.

[112] Vernadsky VI. The Biosphere. New York: Copernicus Springer-Verlag; 1998.
[113] Phillips JD. Self-organization and landscape evolution. Prog. Phys. Geog. 1995;19:309–21.
[114] Krumbein WE, Dyer BD. This planet is alive – weathering and biology, a multi-faceted problem. In: Drever JI, editor. The Chemistry of Weathering. Dordrecht, the Netherlands: D. Reidel; 1985. p. 143–60.
[115] Cambardella, C., S. Logsdon, and D. Olk (2003). Biogeochemical processes influencing soil structure and organic carbon sequestration. NSTL Annual Report, <http://www.nstl.gov/research/ss-ocs.html> accessed February 2007.
[116] Ciaravella A, Scappini F, Franchi M, Cecchi-Pestellini C, Babera M, Candia R, Gallori E, Micela G. Role of clays in protecting adsorbed DNA against X-ray radiation. Int. J. Astrobiol. 2004;3(1):31–5.
[117] Ferris JP, Hill Jr. AR, Liu R, Orgel LE. Synthesis of long prebiotic oligomers on mineral surfaces. Nature 1996;381:59–61.
[118] Laird D. Nature of clay-humic complexes in an agricultural soil: II. Scanning electron microscopy analysis. Soil Sci. Soc. Am. J. 2001;65:1419–25.
[119] Laird DA, Martens DA, Kingery WL. Nature of clay-humic complexes in an agricultural soil, I. Chemical, biochemical and spectroscopic analyses. Soil Sci. Soc. Am. J. 2001;65:1413–8.
[120] Lichtfouse E. A novel model of humin. Analusis 1999;27:385–6.
[121] Lichtfouse E. Temporal pools of individual organic substances in soil. Analusis 1999;27:442–4.
[122] Lichtfouse E. Compound-specific isotope analysis. Application to archaelogy, biomedical sciences, biosynthesis, environment, extraterrestrial chemistry, food science, forensic science, humic substances, microbiology, organic geochemistry, soil science and sport. Rap. Comm. Mass Spectrom. 2000;14:1337–44.
[123] Lichtfouse E, Chenu C, Baudin F. Resistant ultralaminae in soils. Org. Geochem. 1997;25(3/4):263–5.
[124] Lichtfouse E, Chenu C, Baudin F, LeBlond C, Da Silva M, Behar F, et al. A novel pathway of soil organic matter formation by selective preservation of resistant straight-chain biopolymers: chemical and isotope evidence. Org. Geochem. 1998;28(6):411–5.
[125] Lovelock J. (1992). The Evolving Gaia Theory. Paper presented at the United Nations University 9/25, Tokyo, Japan.
[126] Harwood, RR. A history of sustainable agriculture. In: Edwards, CA, Lal, R, Madden, P, Miller, RH, House, G. editors. Sustainable Agricultural Systems. Ankeny, Iowa, USA: Soil and Water Conservation Society; 1990. pp. 3–19.
[127] Bernard C. La Science Experimentale. Paris: J. B. Bailliere and Son; 1878.
[128] Suess E. Die Enstchung der Alpen. Vienna: W. Braunmuller; 1875.
[129] Lotka AJ. The law of evolution as a maximal principle. Hum. Biol. 1945;17:167–94.
[130] Charden PT. The Vision of the Past. London: Collins; 1966.
[131] Lovelock J. Gaia: A New Look at Life on Earth. Oxford, UK: Oxford University Press; 1979.
[132] Westbroek P. Life as a Geological Force: Dynamics of the Earth. New York: Norton Press; 1992.
[133] Flannery T. The Weather Makers: How Man Is Changing the Climate and What It Means for Life on Earth. New York: First Grove Press; 2005.
[134] Sagan D. Notes from the Holocene: A Brief History of the Future. White River Junction, VT: Chelsea Green Publishing Company; 2007.

[135] Pahl-Wostl C. The Dynamic Nature of Ecosystems: Chaos and Order Entwined. New York: John Wiley and Sons; 1995.
[136] Jones B. Diagenetic processes associated with plant roots and microorganisms in karst terrains of the Cayman Islands, British West Indies. In: Wolf KH, Chilingarian GV, editors. Diagenesis, IV. Developments in Sedimentology, vol. 51. Amsterdam: Elsevier; 1994. p. 425–75.
[137] Bruand A, Cousin I, Nicoullaud B, Duval O, Begon JC. Backscattered electron scanning images of soil porosity for analyzing soil compaction around roots. Soil Sci. Soc. Am. J. 1996;60:895–901.
[138] Danin A, Wieder M, Wieder M, Margaritz M. Rhizofossils and root grooves in the Judean Desert and their paleoenvironmental significance. Israel J. Earth Sci. 1987;36:91–9.
[139] Jackson RB, Mooney HA, Schulze ED. A global budget for fine root biomass, surface area, and nutrient contents. Proc. Natl. Acad. Sci. U.S.A 1997;94:7362–6.
[140] Phillips WS. Depth of roots in soil. Ecology 1963;44:424.
[141] Bailey VL, Smith JL, Bolton Jr. H. Fungal-to-bacterial ratios in soils investigated for enhanced C sequestration. Soil Biol. Biochem. 2002;34(7):997–1007.
[142] Dudal Y, Sevenier G, Dupont L, Guillon E. Fate of the metal-binding soluble organic matter throughout a soil profile. Soil Sci. 2005;170(9):707–15.
[143] Walker J, Hays P, Kastink J. A negative feedback mechanism for the long-term stabilization of Earth's surface temperature. J. Geophys. Res. 1981;86:9776–82.
[144] Ward PD, Brownlee D. Rare Earth: Why Complex Life Is Uncommon in the Universe. New York: Copernicus; 2000.
[145] Nugent MA, Brantley SL, Pantano CG, Maurice PA. The influence of natural mineral coatings on feldspar weathering. Nature 1998;395:588–91.
[146] Schopf JW. Cradle of Life: The Discovery of Earth's Earliest Fossils. Princeton, NJ: Princeton University Press; 1999.
[147] Tice MM, Lowe DR. The origin of carbonaceous matter in pre-3.0 Ga greenstone terrains: a review and new evidence from the 3.42 Ga Buck Reef Chert. Earth-Sci. Rev. 2006;76(3–4):259–300.
[148] Oparin AI. The Origin of Life. New York: Macmillan Publishing Co.; 1938.
[149] Brady NC. The Nature and Properties of Soils. New York: MacMillan Publishing Co.; 1974.
[150] Lundqvist M, Nygren P, Jonsson BH, Broo K. Introduction of structure and function in a designed peptide upon absorption on a silica nanoparticle. Angew. Chem. 2006;45:8169–73.
[151] Williams LB, Canfield B, Voglesonger KM, Holloway JR. Organic molecules formed in a "primordial womb." Geology 2005;33(11):913–6.
[152] Schnitzer M. Binding of humic substances by soil mineral colloids. In: Huang PM, Schnitzer M, editors. Interactions of Soil Minerals with Natural Organics and Microbes. Madison, WI: Soil Science Society of America; 1986. p. 78–102. SSSA Special Publication No. 17.
[153] Whitney G, Northrop HR. Experimental investigation of the smectite to illite reaction: dual reaction mechanisms and oxygen-isotope systematics. Am. Mineral. 1988;73:77–90.
[154] Wolfe DW. Tales from the Underground: A Natural History of Subterranean Life. Cambridge, MA: Perseus; 2001.
[155] Cairns-Smith AG. The first organisms. Sci. Am. 1985;252(6):90–100.
[156] Cairns-Smith AG, Hartman H. Clay Minerals and the Origin of Life. Cambridge: Cambridge University Press; 1986.

[157] Rode BM. Peptides and the origin of life. Peptides 1999;20:773–86.
[158] Gobat J-M, Aragno M, Matthey W. The Living Soil. Enfield, NH: Science Publishers, Inc.; 2004.
[159] Gordon SJ, Dorn RI. Localized weathering: implications for theoretical and applied studies. Prof. Geogr. 2005;57(1):28–43.
[160] Stephen I. A study of rock weathering with reference to the soils of the Malvern Hills: part 1. In: Drew JV, editor. Weathering of Biotitite and Granite. Selected Papers in Soil Formation and Classification. Madison, WI: Soil Science Society of America; 1967. p. 311–25.
[161] Buurman P, Jongmans AG. Podzolisation and soil organic matter dynamics. Geoderma 2005;125:71–83.
[162] Bartlett RJ. Oxidation-reduction status of aerobic soils. In: Dowdy RH, editor. Chemistry in the Soil Environment. Madison, WI: American Society of Agronomy; 1981. p. 77–102. ASA Special Publication No. 40.
[163] Bartlett RJ. Soil redox behavior. In: Sparks DL, editor. Soil Physical Chemistry. Boca Raton FL: CRC Press, Inc.; 1986. p. 179–207.
[164] Bartlett RJ, James BR. Redox chemistry of soils. In: Advances in Agronomy. Academic Press, 1993. p. 151–208.
[165] Collins MJ, Bishop AN, Farrimond P. Sorption by mineral surfaces: rebirth of the classical condensation pathway for kerogen formation. Geochim. Cosmochim. Ac. 1995;59(11):2387–91.
[166] Fortin D, Langley S. Formation and occurrence of biogenic iron-rich minerals. Earth-Sci. Rev. 2005;72(1–2):1–19.
[167] Stevenson FJ, Fitch A. Chemistry of complexation of metal ions with soil solution organics. In: Huang PM, Schnitzer M, editors. Interactions of Soil Minerals with Natural Organics and Microbes. Madison, WI: Soil Science Society of America; 1986. p. 29–58. SSSA Special Publication No. 17.
[168] McBride MB. Environmental Chemistry of Soils. New York: Oxford University Press; 1994.
[169] Tan KH. Principles of Soil Chemistry. New York: Marcel Dekker, Inc.; 1998.
[170] Tan KH. Degradation of soil minerals by organic acids. In: Huang PM, Schnitzer M, editors. Interactions of Soil Minerals with Natural Organics and Microbes. Madison, WI: Soil Science Society of America; 1986. p. 1–28. SSSA Special Publication No. 17.
[171] Hayes MHB, Himes FL. Nature and properties of humus-mineral complexes. In: Huang PM, Schnitzer M, editors. Interactions of Soil Minerals with Natural Organics and Microbes. Madison, WI: Soil Science Society of America; 1986. p. 103–58. SSSA Special Publication No. 17.
[172] Huang PM, Violante A. Influence of organic acids on crystallization and surface properties of precipitation products of aluminum. In: Huang PM, Schnitzer M, editors. Interactions of Soil Minerals with Natural Organics and Microbes. Madison, WI: Soil Science Society of America; 1986. p. 159–222. SSSA Special Publication No. 17.
[173] Amundson R. The carbon budget in soils. Ann. Rev. Earth Planet. Sci. 2001;29:535–62.
[174] Frink DS. Explorations into a Dynamic Process-Oriented Soil Science. Ph.D. Dissertation, School of Geographical Sciences. Tempe, Arizona State University; 2007.
[175] Coleman DC, Hunter MD, Hutton J, Pomeroy S, Swift L. Soil respiration from four aggrading forested watersheds measured over a quarter century. Forest Ecol. Manag. 2002;157:247–53.

[176] Hirano T, Kim H, Tanaka Y. Long-term half-hourly measurement of soil CO[2] concentration and soil respiration in a temperate deciduous forest. J. Geophys. Res. 2003;108(D20):4631.
[177] Coyne LM, Lawless J, Lahav N, Sutton S, Sweeney M. Clays as prebiotic photocatalysts. In: Wolman Y, editor. Origins of Life. Hingham, MA: D. Reidel; 1881. p. 115–24.
[178] Velde B, Vasseur G. Estimation of the diagenetic smectite to illite transformation in time-temperature space. Am. Mineral. 1992;77:967–76.
[179] Bartlett RJ. Manganese redox reactions and organic interactions in soils. In: Graham RD, Hannam RJ, Uren NC, editors. Manganese in Soils and Plants. Boston, MA: Kluwer Academic Publishers; 1988. p. 59–73.
[180] DeVrind-DeJong EW, Corstjens PLAM, Kempers ES, Westbroek P, Devrind JPM. Oxidation of manganese and iron by *Leptothrix discophora*: use of N N, N', N'-tetramethyl-p-phenylenediamine as an indicator of metal oxidation. Appl. Environ. Microbiol. 1990;56(11):3458–62.
[181] Douchet FJ, Schneider C, Bones SJ, Kretchmer A, Moss I, Tekely P, Exley C. The formation of hydroxyaluminosilicates of geochemical and biological significance. Geochim. Cosmochim. Ac. 2001;65:2461–7.
[182] Winter Y, Voigt C, Helversen O. Gas exchange during hovering flight in a nectar-feeding bat Glossophaga soricina. J. Exp. Biol. 1998;201:237–44.
[183] Hesse PR. A Textbook of Soil Chemical Analysis. New York: Chemical Publishing Co.; 1971.
[184] Margulis L, Sagan D. Origins of Sex: Three Billion Years of Genetic Recombination. New Haven, CT: Yale University Press; 1986.
[185] Joyce GF. The antiquity of RNA-based evolution. Nature 2002;418:214–21.
[186] Unrau PJ, Bartel DP. RNA-catalysed nucleotide synthesis. Nature 1998;395:260–3.
[187] Rogers J, Joyce GF. A ribozyme that lacks cytidine. Nature 1999;402:323–5.
[188] Montgomery DR. Dirt: The Erosion of Civilizations. Berkeley, CA: University of California Press; 2007.
[189] Arrhenius GO. Crystals and life. Helv. Chim. Acta 2003;85:1569–85.
[190] Bernal JD. The Physical Basis of Life. London: Routledge and Kegan Paul; 1951.
[191] Ferris JP, Joshi PC, Wang KJ, Miyakawa S, Huang W. Catalysis in prebiotic chemistry: application to the synthesis of RNA oligomers. Adv. Space Res. 2004;33:100–5.
[192] Martin W, Russell MJ. On the origins of cells: a hypothesis for the evolutionary transitions from abiotic geochemistry to chemoautotropic prokaryotes, and from prokaryotes to nucleated cells. *Phil. Trans. Roy. Soc. London B* 2003;358:59–85.
[193] Mortland MM. Mechanisms of adsorption of nonhumic organic species by clays. In: Huang PM, Schnitzer M, editors. Interactions of Soil Minerals with Natural Organics and Microbes. Madison, WI: Soil Science Society of America; 1986. p. 59–76. SSSA Special Publication No. 17.
[194] Eglia M, Mirabellab A, Fitzea P. Formation rates of smectites derived from two Holocene chronosequences in the Swiss Alps. Geoderma 2003;117(1–2):81–98.
[195] McMenamin M, McMenamin D. Hypersea: Life on Land. New York: Columbia University Press; 1994.
[196] Stamets P. Mycelium Running: How Mushrooms Can Help Save the World. Berkeley, CA: Ten Speed Press; 2005.
[197] Stamets P. Six ways mushrooms can save the world. Available from: http://www.ted.com/index.php/talks/paul_stamets_on_6_ways_mushrooms_can_save_the_world.html; 2008.

[198] Blum JD, Klaue A, Nezat CA, Driscoll CT, Johnson CE, Siccama TG, Eagar C, Fahey TJ, Likens GE. Mycorrhizal weathering of apatite as an important calcium source in base-poor forest ecosystems. Nature 2002;417:729–31.
[199] Villegas J, Fortin JA. Phosphorus solubilization and pH changes as a result of the interactions between soil bacteria and arbuscular mycorrhizal fungi on a medium containing NH4+ as nitrogen source. Can. J. Bot. 2001;79(8):865–70.
[200] del Val C, Barea JM, Azcon-Aguilar C. Assessing the tolerance to heavy metals of arbuscular mycorrhizal fungi isolated from sewage sludge-contaminated soils. Appl. Soil Ecol. 1999;11(2-3):261–9.
[201] Joner EJ, Briones R, Leyval C. Metal-binding capacity of arbuscular mycorrhizal mycelium. Plant. Soil 2000;226(2):227–34.
[202] Krantz-Rülcker C, Allard B, Schnurer J. Interactions between a soil fungus, *Trichoderma harzianum*, and IIb metals—adsorption to mycelium and production of complexing metabolites. BioMetals 1993;6(4):223–30.
[203] Lamble GM, Nicholson D, Moen A, Berthelsen B, MacDowell AA, Celestre RS, Padmore HA. Zinc speciation in fungus from contaminated forest soils. *Beam Line* 10.3.2 1997:1–2.
[204] Andersson BE, Welindera L, Olssonb PA, Olssonb S, Henryssona T. Growth of inoculated white-rot fungi and their interactions with the bacterial community in soil contaminated with polycyclic aromatic hydrocarbons, as measured by phospholipid fatty acids. Bioresource Technol. 2000;73(1):29–36.
[205] Margulis L, Schwartz KV. Five Kingdoms: An Illustrated Guide to the Phyla of Life on Earth. New York: W. H. Freeman and Company; 1998.
[206] Blackwell M. Terrestrial life – fungal from the start? Science 2000;289:1884–5.
[207] Fisher DW, Bessette AE. Edible Wild Mushrooms of North America: A Field-to-Kitchen Guide. Austin, TX: University of Texas; 1992.
[208] Damasio AR. Descartes' Error: Emotion, Reason, and the Human Brain. New York: Quill; 2000.
[209] Dismukes GC. The metal centers of the photosynthetic oxygen-evolving complex. Photochem. Photobiol. 1986;43:99–115.
[210] Morrison RT, Boyd RN. Organic Chemistry. Boston, MA: Allyn and Bacon, Inc.; 1973.
[211] Lovelock CE, Wright SF, Clark DA, Ruess RW. Soil stocks of glomalin produced by arbuscular mycorrhizal fungi across a tropical rain forest landscape. Smithsonian Environmental Research Center; 2001.
[212] Tornberg K. Wood-decomposing fungi: soil colonization, effects on indigenous bacterial community in soil and hydroxyl radical production. Ph.D. Dissertation, Department of Microbial Ecology, Lund University, Sweden, Lund; 2001
[213] Allen MF, Clouse SD, Weombai BS, Jeakins SL, Friese CF, Allen EB. Mycorrhizae and the integration of scales: from molecules to ecosystems. In: Allen MF, editor. Mycorrhizal Functioning: An Integrative Plant-Fungal Process. New York: Chapman & Hall; 1992. p. 488–515.
[214] Kendrick B. The Fifth Kingdom. Newburyport, MA: Focus Publishing; 2000.
[215] Belnap J. The world at your feet: desert biological soil crusts. Front. Ecol. Environ. 2004;1:181–9.
[216] Belnap J, Lange OL. Structure and functioning of biological soil crusts: a synthesis. In: Belnap J, Lange OL, editors. Biological Soil Crusts: Structure, Function, and Management. Berlin: Springer-Verlag; 2001. p. 471–9. Ecological Studies 150
[217] Hutchens AR. A Handbook of Native American Herbs. London: Shambhala; 1992.

[218] Kavasch EB. American Indian Earth Sense: Herbaria of Ethnobotany and Ethnomycology. Washington, Connecticut: Birdstone Publishers; 1996.
[219] Pettit GR, Tan R, Herald DL, Hamblin J, Pettit RK. Antineoplastic agents. 488. Isolation and structure of yukonin from a Yukon Territory fungus. J. Nat. Prod. 2003;66(2):276–8.
[220] Pettit RK. Antifungal and cancer cell growth inhibitory activites of 1-(3′,4′,5′-trimethoxyphenyl)-2-nitro-ethylene. Mycoses 2002;45:65–74.
[221] Volk, T. (2001). Tom Volk's Fungus of the Month for December 2001. Available from: http://botit.botany.wisc.edu/toms_fungi/dec2001.html.
[222] Barea JM, Azcón R, Azcon-Aguilar C. Mycorrhizosphere interactions to improve plant fitness and soil quality. Anton. Leeuw. 2002;81(1-4):343–51.
[223] Fitter AH, Garbaye J. Interactions between mycorrhizal fungi and other soil organisms. Plant. Soil. 1994;159(1):123–32.
[224] Currie CR, Wong B, Stuart AE, Schultz TR, Rehner SA, Mueller UG, Sung GH, Spatafora JW, Straus NA. Ancient tripartite coevolution in the attine ant-microbe symbiosis. Science 2003;299(5605):386–8.
[225] Boynton, D. (1941). Soils in relation to fruit growing in New York. Part XV. Cornell University Agricultural Experiment Station.
[226] Stolzy LH, Letey J, Szuskiewicz TE, Lunt OR. Root growth and diffusion rates as functions of oxygen concentration. Soil Sci. Soc. Am. Proc. 1961;2:463–7.
[227] Van't Hoff JH. Studies of Chemical Dynamics. Amsterdam: Frederik Muller and Co; 1884.
[228] Dalal RC. Acidic soil pH, aluminum and iron affect organic carbon turnover in soil. In: Kirschbaum MUF, Mueller R, editors. Net Ecosystem Exchange: Workshop Proceedings. Canberra, Australia: CRC for Greenhouse Accounting; 2001. p. 111–5.
[229] Simonson RW. Outline of a generalized theory of soil genesis. Soil Sci. Soc. Am. Proc. 1959;23:152–6.
[230] Frink DS. The chemical variability of carbonized organic matter through time. Archaeol. E. N. Am. 1992;20:67–79.
[231] Dowman EA. Conservation in Field Archaeology. London: Methuen; 1970.
[232] Cohen-Ofri, Ilit, Lev Weiner, Elisabetta Boaretto, Genia Mintz, and Steve Weiner. Modern and fossil charcoal: aspects of structure and diagenesis. J. Archaeol. Sci. 2006; 33:428–39.
[233] Frink DS. The oxidizable carbon ratio (OCR): a proposed solution to some of the problems encountered with radiocarbon data. N. Am. Archaeol. 1994;15(1):17–29.
[234] Ruffner J. Climates of the States: National Oceanic and Atmospheric Administration. Detroit, MI: Gale Research Co; 1978.
[235] Chandler RF. A study of certain calcium relationships and base exchange properties of forest soils. J. Forest. 1937;35:27–32.
[236] Hissink DJ. The reclamation of the Dutch saline soils (Solonhak) and their further weathering under the humid climatic conditions of Holland. Soil Sci. 1938;45:83–94.
[237] Salisbury EJ. Notes on the edaphic succession in some dune soils with special reference to the time factor. J. Ecol. 1925;13:322–8.
[238] Buol SW. Present soil-forming factors and processes in arid and semiarid regions. Soil Sci. 1965;99:45–9.
[239] Hole FD. An ancient young soil. Soil Survey Hor. 1967;8:16–19.
[240] Scharpenseel HW. Radiocarbon dating of soils: problems, troubles, hopes. In: Yaalon DH, editor. Paleopedology: Origin, Nature and Dating of Paleosols. Jerusalem: ISSS and Israel Univ. Press; 1971. p. 77–88.

[241] Gillespie R, Prosser IP, et al. AMS dating of alluvial sediments on the southern tablelands of New South Wales, Australia. Radiocarbon 1992;34:29–36.

[242] deVries, H. (1958). Variations in concentration of radiocarbon with time and location on Earth. Nederlandsche Akademie van Wetenschappen.

[243] Suess H. Radiocarbon concentrations in modern wood. Science 1955;120:5–7.

[244] Rafter T, Ferguson G. The atom bomb effect. Recent increase in the C-14 content of the atmosphere, biosphere, and surface waters of the oceans. New Zeal. J. Sci. Technol. 1957;B38:871–83.

[245] Ralph E. Carbon-14 dating. In: Brill R, editor. Dating Techniques for the Archaeologist. Cambridge, MA: MIT Press; 1971. p. 244–50.

[246] Markewich HW, Pavich MJ, Maushbach MJ, Johnson RG, Gonzalez VM. A Guide for Using Soil and Weathering Profile Data in Chronosequence Studies of the Coastal Plain of the Eastern United States. Washington, D.C.: U.S. Geological Survey; 1989.

[247] Schaetzl RJ, Barrett R, Winkler JA. Choosing models for soil chronofunctions and fitting them to data. Eur. J. Soil Sci. 1994;45:219–32.

[248] Yaalon DH. Conceptual models in pedogenesis: can soil-forming functions be solved? Geoderma 1975;14:189–205.

[249] Frink DS. Application of the OCR dating procedure, and its implications for pedogenic research. In: Collins ME, editor. Pedological Perspectives in Archaeological Research. Madison, WI: Soil Science Society of America; 1995. p. 95–106.

[250] Harrison R, Frink D. The OCR carbon dating procedure in Australia: new dates from Wilinyjibari Rockshelter, Southeast Kimberly, Western Australia. Aust. Archaeol. 2000;51:6–15.

[251] Nami HG, Frink D. Chronologia obtenida pro la tasa del carbono organico oxidable (OCR) en Markatch Aike 1 (Cuenca Del Rio Chico, Santa Cruz). An. Instit. Patag. 1999;27:231–7.

[252] Saunders JW, Mandel RD, Saucier RT, Allen ET, Hallmark CT, Johnson JK, et al. A mound complex in Louisiana at 5400-5000 years before the present. Science 1997;277:1796–9.

[253] Frink DS. Temporal values in a universe of turbations: application of the OCR carbon dating procedure in archaeological site formational analyses and pedogenic evaluations. In: Fuleky G, editor. Soils and Archaeology: Papers of the 1st International Conference on Soils and Archaeology, Szazhalombatta, Hungary 30 May – 3 June 2001. Oxford, UK: Archaeopress; 2003. p. 5–12.

[254] Frink DS. Examination of the unexplored balk between pedology and archaeology. In: Boschian G, editor. Proc. 2nd international conference on soils and archaeology. Pisa: Dipartimento di Scienze Archeologiche, Universita di Pisa; 2003. p. 31–3.

[255] Frink DS. (n.d.). Taphonomic Processes Affecting Monumental Earthen Architecture as a Proxy for Climatic Change. In: Comparative Archeology and Paleoclimatology: Sociolcultural Responses to a Changing World (M. Baldia, T. Perttula, and D. S. Frink, editors.), London, UCL Press.

[256] Frink DS, Dorn RI. Beyond taphonomy: pedogenic transformations of the archaeological record in monumental earthworks. J. Arizona-Nevada. Acad. Sci. 2002;4(1):24–44.

[257] Martinez G. A preliminary report of the Late Pleistocene site of Paso Otero 5 in the Pampean Region of Argentina. *Curr. Res. Pleistocene* 1997;14:53–5.

[258] Holliday V, Martinez G, Johnson E, Buchanan B. Geoarchaeology of Paso Otero 5 (Pampas of Argentina). In: Miotti L, Salemme M, Flegenheimer N, editors. Ancient Evidences for Paleo South Americans: From Where the South Winds Blow: Center

for the Studies of the First Americans (CSFA) and Texas A&M University Press; 2003. p. 37–43.
[259] Visual Comput. 1985;1:69–97.
[260] Edsall RM. The parallel coordinate plot in action: design and use in geographic visualization. Comput. Stat. Data An. 2003;43:605–19.
[261] Phillips JD. Divergence, convergence, and self-organization in landscapes. Ann. Assoc. Am. Geogr. 1999;89(3):466–88.
[262] Phillips JD. Contingency and generalization in pedology, as exemplified by texture-contrast soils. Geoderma 2001;102(3-4):347–70.
[263] Robinson AR. Sediment yield as a function of upstream erosion. In: Peterson AE, Swan JB, editors. Universal Soil Loss Equation: Past, Present, and Future. Madison, WI: Soil Science Society of America; 1978. p. 7–16. SSSA Special Publication No. 8.
[264] Bird M, Santruckova H, Lloyd J, Lawson E. The isotopic composition of soil organic carbon on a north-south transect in western Canada. Eur. J. Soil Sci. 2002;53(3):393–403.
[265] Hagedorn F, Spinnler D, Siegwolf R. Increased N deposition retards mineralization of old soil organic matter. Soil Biol. Biochem. 2003;35:1683–92.
[266] Harris WG, Crownover SH, Hinchee J. Problems arising from fixed-depth assessment of deeply weathered sandy soils. Geoderma 2005;126:164–5.
[267] Jobbagy EG, Jackson RB. The vertical distribution of soil organic carbon and its relation to climate and vegetation. Ecol. Appl. 2000;10(2):423–36.
[268] Rumpel C, Kogel-Knabner I. Microbial use of lignite compared to recent plant litter as substrates in reclaimed coal mine soils. Soil Biol. Biochem. 2004;36:67–75.
[269] Schmidt MWI, Kogel-Knabner I. Organic matter in particle-size fractions from A and B horizons of a haplic alisol. Eur. J. Soil Sci. 2002;52:383–92.
[270] Fang C, Smith P, Moncrieff JB, Smith JU. Similar response of labile and resistant soil organic matter pools to changes in temperature. Nature 2005;433:57–9.
[271] Fritsen CH, Grue AM, Priscu JC. Distribution of organic carbon and nitrogen in surface soils in the McMurdo Dry Valleys, Antarctica. Polar Biol. 2000;23:121–8.
[272] Mayorga E, Aufdenkampe AK, et al. Young organic matter as a source of carbon dioxide outgassing from Amazonian rivers. Nature 2005;436:538–41.
[273] Moore JC, McCann K, De Ruiter PC. Modeling trophic pathways, nutrient cycling and dynamic stability in soils. Pedobiologia 2005;49(6):499–510.
[274] Wallander H, Nilsson LO, Hagerberg D, Baath E. Estimation of the biomass and seasonal growth of external mycelium of ectomycorrhizal fungi in the field. New Phytol. 2001;151(3):753.
[275] Killham K. Soil Ecology. Cambridge: Cambridge University Press; 1994.
[276] Collins BS, Dunne KP, Pickett STA. Responses of forest herbs to canopy gaps. In: Pickett STA, White PS, editors. The Ecology of Natural Disturbance and Patch Dynamics. Orlando, FL: Academic Press; 1985. p. 218–34.
[277] Six J, Conant RT, Paul EA, Paustian K. Stabilization mechanisms of soil organic matter: implications for C-saturation of soils. Plant. Soil. 2002;241:115–76.
[278] Jones D, Muehlchen A. Effects of the potentially toxic metals, aluminium, zinc and copper on ectomycorrhizal fungi. J. Environ. Sci. Heal. A. 1994;29:949–66.
[279] Moore JM, Picker MD. Heuweltjies (earth mounds) in the Clanwillian district, Cape Province South Africa: 4000-year-old termite nests. Oecologia 1991;86(3):424–62.
[280] Geosci. Man 1978;19:23–40.
[281] Obermeier SF. Use of liquefaction-induced features for paleoseismic analysis – an overview of how seismic liquefaction features can be distinguished from other features and

how their regional distribution and properties of source sediment can be used to infer the location and strength of Holocene paleo-earthquakes. Eng. Geol. 1996;44(1):1–76.
[282] Flint RF. Glacial and Quaternary Geology. New York: John Wiley & Sons.; 1971.
[283] Perttula TK. The Caddo Nation: Archaeological and Ethnohistoric Perspectives. Austin, TX: University of Texas Press; 1992.
[284] Maiuri A. Pompeii. Rome: Institute Poligrafico dello Stato; 1970.
[285] Cornwall IW. Soils for the Archaeologist. London: Phoenix House; 1958.
[286] Holliday VT, editor. Soils in Archaeology: Landscape Evolution and Human Occupation. Washington, D.C: Smithsonian Institution Press; 1992.
[287] Limbrey S. Science in Archaeology. London: Academic Press; 1975.
[288] Stewart RN, Bailey, Oghenekome O. Soil science for archaeologists. Florida Agricultural and Mechanical University Southeast Archaeological Center, National Park Service; 2000.
[289] Vogel G. A Handbook of Soil Description for Archeologists. Fayetteville, AR: Arkansas Archeological Survey; 2002.
[290] Courty MA, Goldgerg P, MacPhail R. Soils and Micromorphology in Archaeology. New York: Cambridge University Press; 1989.
[291] Cantley CE, Loubser JHN, Vojnovski P, Langdale J, Young B. Cultural Resources Survey and Site Evaluation: Ten Mile Creek Water Preserve Area Critical Project, St. Lucie County, FL. Jacksonville, FL. U.S. Army Corps of Engineers: 114; 2003.
[292] Polyak VJ, Asmerom Y. Late Holocene climate and cultural changes in the Southwestern United States. Science 2001;294:148–51.
[293] Wurster CM, Patterson WP. Late Holocene climate change for the eastern interior United States: evidence from high-resolution *180 values of sagittal otoliths. Palaeogeogr. Palaeocl. 2001;170:81–100.
[294] Brown SL, Bierman PR, Lini A, Southon J. 10,000 yr record of extreme hydrologic events. Geology 2000;28(4):335–8.
[295] Gunn JD, Foss JE, Folan WJ, Carrasco M dRD, Faust BB. Bajo sediments and the hydraulic system of Calakmul, Campeche, Mexico. Anc. Mesoam. 2002;13:297–315.
[296] Hodell DA, Brenner M, Curtis J, Guilderson T. Solar forcing of drought frequency in the Maya Lowlands. Science 2001;292:1367–70.
[297] Schimmelmann A, Lange CB, Meggers BJ. Palaeoclimatic and archaeological evidence for a ~200-yr recurrence of floods and droughts linking California, Mesoamerica and South America over the past 2000 years. Holocene 2003;13(5):763–78.
[298] Caviedes CN. El Nino in History: Storming through the Ages. Gainesville, FL: University of Florida; 2001.
[299] Hughes MJ, Lampert RJ. Occupational disturbance and types of archaeological deposits. J. Archaeol. Sci. 1977;4:135–40.
[300] Gunn JD, editor. BAR International Series. Oxford: Archaeopress; 2000.
[301] Smíd M. Výsledky zjist'ovacího výzkumu na eneolitickém hradisku Rmíz u Laskova. Pravek. Nova Rada. 1995;3:19–77.
[302] Baldia MO. The oldest stone rampart: enclosures and megalithic tombs of the Funnel Beaker Culture (4100-2800 cal BC) in Central Europe. In: Jadin I, Hauzeur A, Cauwe N, Vander Linde M, Tunca Ö, Lebeau M, editors. Section 9. Le Néolithique au Proche Orient et en Europe/The Neolithic in the Near East and Europe. Section 10: L'âge du cuivre au Proche Orient et en Europe/The Copper Age in the Near East and Europe. Oxford: Archaeopress; 2004. p. 153–61. Actes du XIVème Congrès UISPP, Universitè de Liège, Belgique, 2-8 Septembre 2001. BAR International Series 1303
[303] Baldia MO, Boulanger MT, Frink DS. The earthen long-barrow of Dzban, Moravia, Czech Republic and its implications for the interaction between the Nordic funnel

beaker and the southern Baden culture. In: The Baden Complex and the Outside World (M. Furholt, M. Szmyt, and A. Zastawny, editors., in coop. with E. Schalk), Proceedings of the twenth annual meeting of the EAA in Cracow, September 19–24, 2006. Studien zur Archäologie in Ostmitteleuropa, Habelt, Bonn; 2008. p. 263–287.

[304] Baldia MO, Frink DS, Boulanger MT. Problems in the archaeological legacy: the TRB/Lengyel-Baden conundrum. In: The Baden Complex and the Outside World (M. Furholt, M. Szmyt, and A. Zastawny, editors., in coop. with E. Schalk), Proceedings of the twelfth annual meeting of the EAA in Cracow, September 19–24, 2006. Studien zur Archäologie in Ostmitteleuropa, Habelt, Bonn; 2008. p. 25–48.

[305] Honeisen M. Die Ausbreitung fruhster bauerlicher Kultur in Europa. In: Honeisen M, editor. Die Ersten Bauern 2: Einfuhrung, Balkan, andrenzende Regionen der Schweiz. Zurich: Schweizerisches Landesmuseum; 1990. p. 15–26.

[306] Nicholas GP. Ecological leveling: the archaeology and environmental dynamics of early post glacial land use. In: Nicholas GP, editor. Holocene Human Ecology in Northeastern North America. New York: Plenum Pres; 1988. p. 257–96.

[307] Frink DS, Hathaway AD. Behavioral continuity on a changing landscape. In: Cremeens DL, Hart JP, editors. Geoarchaeology of Landscapes in the Glaciated Northeast: Proceedings of a Symposium held at the New York Natural History Conference VI. Albany, NY: The New York State Museum; 2003. p. 103–16.

[308] Jennings KL, Bierman PR, Southon J. Timing and style of deposition on humid-temperate fans, Vermont, United States. Geol. Soc. Am. Bull. 2003;115(2):182–99.

[309] Johnsen SJ, Dahl-Jensen D, Gundestrup N, Steffensen JP, Clausen HB, Miller H, Masson-Delmotte V, Sveinbjornsdottir AE, White J. Oxygen isotope and palaeotemperature records from six Greenland ice-core stations: Camp Century, Dye-3, GRIP, GISP2, Greenland and NorthGRIP. J. Quaternary Sci. 2001;16(4):299–307.

[310] Alley RB, Meese DA, Shuman CA, Gow AJ, Taylor KC, Grootes PM, White JM, Ram M, Waddington ED, Mayewski PA, Zielinski GA. Abrupt increases in Greenland snow accumulation at the end of the Younger-Dryas event. Nature 1993;362:527–9.

[311] Dansgaard W, White JWC, Johnsen SJ. The abrupt termination of the Younger Dryas climate event. Nature 1989;339:532–3.

[312] Taylor KC, Lamorey GW, Doyle GA, Alley RB, Grootes PM, Mayewski PA, White JW, Barlowet LK. The "flickering switch" of Late Pleistocene climate change. Nature 1993;361:432–6.

[313] Haviland W, Power MW. The Original Vermonters: Native Inhabitants, Past and Present. Hanover: University Press of New England; 1981.

[314] Loring S. Paleo-Indian hunters and the Champlain Sea: a presumed association. Man. Northeast. 1980;19:15–41.

[315] Anderson DG, Gillam JC. Paleoindian colonization of the Americas: implications from an examination of physiography, demography, and artifact distribution. Am. Antiquity 2000;65:43–66.

[316] Watkins J. Indigenous Archaeology: American Indian Values and Scientific Practice. New York: Alta Mira Press; 2000.

[317] Clark CC, Custer JF. Rethinking Delaware archaeology: a beginning. N. Am. Archaeol. 2003;24(1):29–81.

[318] Morehead JR, Mathews JH, Campbell LJ, Williams CA, Bourgeois P, Thomas PM, Jr. Fort Polk 56: the results of the fifty-sixth program of site testing at Ten Sites, Fort Polk Military Reservation, Vernon Parish, Louisiana. Prentice Thomas and Associates, Inc., Report of Investigations 774. Fort Walton Beach, FL; 2005.

[319] Koestler A. Janus: A Summing Up. New York: Random House; 1978.

[320] Taylor WH, Hotz CF. Soil Survey of Worcester County, Northeastern Part. Washington, D.C.: United States Department of Agriculture, Soil Conservation Service; 1985.
[321] Bellamy PH, Loveland PJ, Bradley RI, Lark RM, Kirk GJD. Carbon losses from all soils across England and Wales 1978-2003. Nature 2005;437:245–8.
[322] Capalbo S, Antle H, Mooney S, Paustian K. Sensitivity of carbon sequestration costs to economic and biological uncertainties. Environ. Manag. 2004;33:238–51.
[323] Carter MR. Analysis of soil organic matter storage in agroecosystems. In: Carter MR, Stewart BA, editors. Structure and Organic Matter Storage in Agricultural Soils. Advances in Soil Science. : CRC Press; 1996. p. 3–8.
[324] Christensen BT, Johnston AE. Soil organic matter and soil quality – lessons learned from long-term experiments at Askov and Rothamsted. In: Gregorich EG, Carter MR, editors. Soil Quality for Crop Production and Ecosystem Health. Amsterdam: Elsevier Science; 1997. p. 399–430.
[325] Copard Y, Di-Giovanni C, Martaud T, Alberic P, Olivier J-E. Using Rock-Eval 6 pyrolysis for tracking fossil organic carbon in modern environments: implications for the roles of erosion and weathering. Earth Surf. Proc. Land. 2006;31:135–53.
[326] Corti G, Ugolini FC, Agnelli A, Certini G, Cunliglio R, Berna F, Fernánez Sanjurjo MJ. The soil skeleton, a forgotten pool of carbon and nitrogen in the soil. Eur. J. Soil Sci. 2002;53:283–98.
[327] Davidson EA, Trumbore SE, Amundson R. Biogeochemistry: soil warming and organic carbon content. Nature 2000;408:789–90.
[328] Falloon P, Smith P. Accounting for changes in soil carbon under the Kyoto Protocol: need for improved long-term data sets to reduce uncertainty in model projections. Soil Use Manage. 2003;19(3):265–9.
[329] Foster RC. Polysaccharides in soil fabrics. Science 1981;214:665–7.
[330] Gleixner G, Bol R, Balesdent J. Molecular insight into soil carbon turnover. Rap. Comm. Mass Spectrom. 1999;13:1278–83.
[331] Johnson D, Krsek M, Wellington EMH, Scott AW, Cole L, Bardgett RD, Read DJ, Leake JR. Soil invertebrates disrupt carbon flow through fungal networks. Science 2005;309:1047.
[332] Kerner M, Hohemberg H, Erti S, Reckermann M, Spitzy A. Self-organization of dissolved organic matter to micelle-like microparticles in river water. Nature 2003;422:150–4.
[333] Knorr W, Prentice LC, House JI, Holland EA. Long-term sensitivity of soil carbon turnover to warming. Nature 2005;433:298–301.
[334] Lal R. Soil carbon sequestration impacts on global climate change and food security. Science 2004;304:1623–7.
[335] Lal R. Soil carbon sequestration to mitigate climate change. Geoderma 2004;123:1–22.
[336] Lal R, Blum WH, Valentin C, Stewart BA, editors. Methods for Assessment of Soil Degradation: Advances in Soil Science. Boca Raton, FL: CRC Press; 1997.
[337] Lal R, Kimble JM, Follett RF. Methodological challenges toward balancing soil C pools and fluxes. In: Lal R, Kimble JM, Follett RF, Stewart BA, editors. Assessment Methods for Soil Carbon. New York: Lewis Publishers; 2001. p. 659–67.
[338] Lal R, Kimble JM, Follett RF, Stewart AA. Assessment Methods for Soil Carbon: Advances in Soil Science. New York: Lewis Publishers; 2001.
[339] Oren R, Ellsworth DS, Johnsen KH, Phillips N, Ewers BE, Maier C, Schäfer KV, McCarthy H, Hendrey G, McNulty SG, Katul GG. Soil fertility limits carbon sequestration by forest ecosystems in a CO2-enriched atmosphere. Nature 2001;411:469–72.
[340] Pawlson D. Will soil amplify climate change? Nature 2005;433:204–5.

[341] Petersen BM, Berntsen J, Hansen S, Jensen LS. CN-SIN – a model for the turnover of soil organic matter. I. Long-term carbon and radiocarbon development. Soil Biol. Biochem. 2005;37:359–74.

[342] Petersen BM, Jensen LS, Hansen S, Pedersen A, Henriksen TM, Sorensen P, Trinsoutrot-Gattin I, Berntsen J. CN-SIM – a model for the turnover of soil organic matter. II. Short-term carbon and nitrogen development. Soil. Biol. Biochem. 2005;37:375–93.

[343] Richter DD, Markewitz D, Wells CG, Allen HL, Dunscomb J, Harrison K, Heine PR, Stuanes A, Urrego B, Bonani G. Carbon cycling in an old-field pine forest: implications for the missing carbon sink and for the concept of soil. In: McFee WW, Kelly JM, editors. Carbon Forms and Functions in Forest Soils. Madison, WI: Soil Science Society of America; 1995. p. 233–51.

[344] Rühlmann J. A new approach to estimating the pool of stable organic matter in soil using data from long-term field experiments. Plant. Soil. 1999;213:149–60.

[345] Schimel D, Melillo J, Tian H, McGuire AD, Kicklighter D, Kittel T, Tosenbloom N, Running S, Thornton P, Ojima D, Parton W, Kelly R, Sykes M, Neilson R, Rizzo B. Contribution of increasing CO_2 and climate to carbon storage by ecosystems in the United States. Science 2000;287:2004–6.

[346] Schlesinger WH, Lichter J. Limited carbon storage in soil and litter of experimental forest plots under increased atmospheric CO_2. Nature 2001;411:466–9.

[347] Simbahan GC, Dobermann A. Sampling optimization based on secondary information and its utilization in soil carbon mapping. Geoderma 2006;133:345–62.

[348] Simbahan GC, Dobermann A, Goovaerts P, Ping J, Haddix ML. Fine-resolution mapping of soil organic carbon based on multivariate secondary data. Geoderma 2006;132:471–89.

[349] Sunda WG, Kieber DJ. Oxidation of humic substances by manganese oxides yields low-molecular-weight organic substrates. Nature 1994;367:62–4.

[350] Theng BKG, Tate KR, Becker-Heidmann P. Towards establishing the age, location, and identity of the inert soil organic matter of a Spodoso. Z. Pflanz. Bodenkunde 1992;155:181–4.

[351] Torn MS, Trumbore SE, Chadwick OA, Vitousek PM, Hendricks DM. Mineral control of soil organic carbon storage and turnover. Nature 1997;389:170–3.

[352] Townsend AR, Vitousek PM, Trumbore SE. Soil organic matter dynamics along gradients in temperature and land use on the island of Hawaii. Ecology 1995;76:721–33.

[353] Treseder KK, Allen MF. Mycorrhizal fungi have a potential role in soil carbon storage under elevated CO_2 and nitrogen deposition. New Phytol. 2000;147:189–200.

[354] Trumbore SE. Potential response of soil organic carbon to global environmental change. Proc. Natl. Acad. Sci. India 1997;94:8284–91.

[355] Vagen TG, Lal R, Singh BR. Soil carbon sequestration in Sub-Saharan Africa: a review. Land Degrad. Dev. 2005;16:53–71.

[356] Wiseman CLS, Puttmann W. Interactions between mineral phases in the preservation of soil organic matter. Geoderma 2006;134:109–18.

[357] Zhou G, Liu S, Li Z, Zhang D, Tang X, Zhou C, Yan J, Mo J. Old-growth forests can accumulate carbon in soils. Science 2006;314:1417.

[358] Barr S. Strategies for sustainability: citizens & responsible environmental behavior. Area 2003;53:227–40.

[359] Barriosa E, Delveb RJ, Bekundac M, Mowod J, Agundae J, Ramischf J, Trejog MT, Thomasa RJ. Indicators of soil quality: a South–South development of a methodological guide for linking local and technical knowledge. Geoderma 2006;135:248–59.

[360] Doran JW. Soil health and global sustainability: translating science into practice. Agr. Ecosyst. Environ. 2002;88:119–27.
[361] Doran JW, Parkin TB. Defining and Assessing Soil Quality. In: Doran JW, Coleman DC, Bezdicek DF, Stewart BA, editors. SSSA Special Publication 35. Madison, WI: American Society of Agronomy; 1994. p. 3–21.
[362] Doran JW, Sarrantonio M, Jaure R. Strategies to promote soil quality and health. In: Pankhurst CE, Doube BM, Gupta VVSR, Grace PR, editors. Soil Biota: Management in Sustainable Farming Systems: Melbourne. : CSIRO; 1994. p. 230–7.
[363] Doran JW, Stamatiadis SI, Haberern J. Soil health as an indicator of sustainable management. Agr. Ecosyst. Environ. 2002;88:107–10.
[364] Haberern J. Viewpoint: a soil health index. J. Soil Water Conserv 1992;47:6.
[365] Hund-Rinke K, Kordel W, Terytze K. Assessment of soil quality: state-of-the-art, Germany. J. Soil. Sediment 2003;3:234.
[366] Lal R. Soil carbon sequestration impacts on global climate change and food security. Science 2004;304:1623–7.
[367] Larson WE, Pierce FJ. The Dynamics of Soil Quality as a Measure of Sustainable Management. In: Doran JW, Coleman DC, Bezdicek DF, Stewart BA, editors. SSSA Special Publication 35. Madison, WI: American Society of Agronomy; 1994. p. 37–51.
[368] Melitz PJ. The sufficiency concept in land evaluation. Soil Surv. Land Eval. 1986;6:9–19.
[369] O'Connell DA, Ryan PJ, McKenzie NJ, Ringrose-Voase AJ. Quantitative site and soil descriptors to improve the utility of forest soil surveys. Forest Ecol. Manag. 2000;138:107–22.
[370] Pankhurst CE. Biological indicators of soil health and sustainable productivity. In: Greenland DJ, Szabolics I, editors. Soil Resilience and Sustainable Land Use. Wallingford, England: CAB International; 1994. p. 331–51.
[371] Pierce FJ, Larson WE, Dowdy RH, Graham WAP. Productivity of soils: assessing long-term changes due to erosion. J. Soil Water Conserv. 1983;38:39–44.
[372] Qi J, Chehbouni A, Huete AR, Kerr YH, Sorooshian S. A modified soil adjusted vegetation index. Remote Sens. Environ. 1994;48:119–26.
[373] Sanchez PA, Palm CA, Buol SW. Fertility capability soil classification: a tool to help assess soil quality in the tropics. Geoderma 2003;114:157–85.
[374] Seryi AI. Theoretical and methodological aspects of soil rating. Eurasian Soil Sci. 1996;28:84–96.
[375] Stewart GA. Land evaluation. In: Stewart GA, editor. Land Evaluation: Papers of a CSIRO Symposium, Organized in Cooperation with UNESCO, Canberra 26–31 August 1968. South Melbourne: Macmillan Company of Australia; 1968. p. 1–10.
[376] USDA Natural Resources Conservation Service (1996). Soil Quality – Introduction. Available at: http://www.statlab.iastate.edu/survey/SQI/pdf/sq_1_one.pdf.
[377] Wall DH, Virginia RA. Controls on soil biodiversity: insights from extreme environments. Applied Soil Ecol. 1999;13:137–50.
[378] Entry JA, Watrud LS, Reeves M. Accumulation of Cs-137 and Sr-90 from contaminated soil by three grass species inoculated with mycorrhizal fungi. Environ. Poll. 1998;104:449–57.
[379] Fresquez PR, Aldon EF, Lindemann WC. Microbial re-establishment and the diversity of fungal genera in reclaimed coal mine spoils and soils. Reclam. Reveg. Res. 1986;4:245–58.
[380] Garcia MA, Alonso J, Fernandez MI, Melgar MJ. Lead content in edible wild mushrooms in northwest Spain as indicator of environmental contamination. Arch. Environ. Contam. Toxicol. 1998;34:330–5.

[381] Khan AG, Chaudhry TM, Hayes WJ, Khoo CS, Hill L, Fernandez R, Gallardo P. Physical, chemical and biological characterisation of a steelworks waste site at Port Kembla, NSW Australia. Water Air Soil Poll. 1998;104:389–402.

[382] Melgar MJ, Alonso J, PerezLopez M, Garcia MA. Influence of some factors in toxicity and accumulation of cadmium from edible wild macrofungi in NW Spain. J. Environ. Sci. Heal. B 1998;33:439–55.

[383] Moller A, Muller HW, Abdullah A, Abdelgawad G, Utermann J. Urban soil pollution in Damascus, Syria: concentrations and patterns of heavy metals in the soils of the Damascus Ghouta. Geoderma 2005;124:63–71.

[384] Munro LJA, Penning-Rowsell EC, Barnes HR, Fordham MH, Jarrett D. Infant mortality and soil type: a case study in south-central England (with discussion). Eur. J. Soil Sci. 1997;48:1–17.

[385] Murray R, Phillips P, Bender J. Degradation of pesticides applied to banana farm soil: comparison of indigenous bacteria and a microbial mat. Environ. Toxicol. Chem. 1997;16:84–90.

[386] Nofziger DL, Chen J-S, Hornsby AG. Uncertainty in pesticide leaching risk due to soil variability. In: Nettleton WD, editor. Data Reliability and Risk Assessment: Applicability to Soil Interpretations. SSSA Special Publication 47. Madison, WI: American Society of Agronomy; 1996. p. 99–130.

[387] Rikken MGJ, Van Rijn RPG. Soil Pollution with Heavy Metals — An Inquiry into Spatial Variation, Cost of Mapping and the Risk Evaluation of Copper, Cadmium, Lead and Zinc in the Floodplains of the Meuse West of Stein, the Netherlands. Utrecht: Dept. of Physical Geography, Utrecht University; 1993.

[388] Rundel PW, Kogel-Knabner I. Microbial use of lignite compared to recent plant litter as substrates in reclaimed coal mine soils. Soil Biol. Biochem. 2004;36:67–75.

[389] Schmidt MWI. Organic matter in natural soils and in soils contaminated by atmospheric organic particles from coal processing industries. Ph.D. dissertation. Shaker Verlag, Aachen, Reihe Geowissenschaften; 1998.

[390] Schmidt MWI, Knicker H, Hatcher PG, Kögel-Knabner I. Airborne contamination of forest soils by carbonaceous particles from industrial coal processing. J. of Environ. Qual. 2000;29:768–77.

[391] Sipos P, Nemeth T, Mohai I, Dodony I. Effect of soil composition on adsorption of lead as reflected by a study on a natural forest soil profile. Geoderma 2005;124:363–74.

[392] Sodergren A. Uptake and accumulation of CH-DDT by *Chlorella* sp. (Chlorophyceae). Oikos 1968;19:126–38.

[393] Steenhuis T, Vandenheuvel K, Weiler KW, Boll J, Daliparthy J, Herbert S, Kung KJS. Mapping and interpreting soil textural layers to assess agrichemical movement at several scales along the eastern seaboard (USA). Nutr. Cycl. Agroecosys. 1998;50:91–7.

[394] Stolpe NB, Kuzila MS, Shea PJ. Importance of soil map detail in predicting pesticide mobility in terrace soils. Soil Sci. 1998;163:394–403.

[395] Tsay SF, Lee JM, Lynd JQ. The interactions of Cu^{++} and Cn^- with paraquat phytotoxicity to a Chorella. Weed Sci. 1970;18:596–8.

[396] Vaishampayan A, Hemantaranjan A. Physiologial significance of vanadium uptake during N_2 and NO_3^- metabolism in various strains of a N_2-fixing cyanobacterium *Nostoc muscorum*. Plant Cell Physiol. 1984;25:845–50.

[397] Ahuja LR. Quantifying agricultural management effects on soil properties and processes. Geoderma 2003;116:1–2.

[398] Bakker JA. The TRB West Group: studies in the chronology and geography of the Makers of Hunebeds and Tiefstich Pottery. Subfaculteit der Pre- en Protohistorie, Amsterdam, Universiteit van Amsterdam; 1979.

[399] Beavis J, Mott CJB. Effects of land use on the amino acid composition of soils: 1. Manured and unmanured soils from the Broadbalk continuous wheat experiment, Rothamsted, England. Geoderma 1996;72:259–70.
[400] Capriel P. Hydrophobicity of organic matter in arable soils: influence of management. Eur. J. Soil Sci. 1997;48:457–62.
[401] Cools N, De Pauw E, Deckers J. Towards an integration of conventional land evaluation methods and farmers' soil suitability assessment: a case study in northwestern Syria. Agr. Ecosyst. Environ. 2003;95:327–42.
[402] Flores S, Saxena S, Stotzky G. Transgenic Bt plants decompose less in soil than non-Bt plants. Soil Biol. Biochem. 2005;37:1073–82.
[403] Fründ R, Haider K, Lüdemann H-D. Impact of soil management practices on the organic matter structure - investigations by CPMAS 13C NMR-spectroscopy. Z. Pflanz. Bodenkunde 1994;157:29–35.
[404] Janzen HH, Campbell CA, Ellert BH, Bremer E. Soil organic matter dynamics and their relationship to soil quality. In: Gregorich EG, Carter MR, editors. Soil Quality for Crop Production and Ecosystem Health. Developments in Soil Science 25. Amsterdam: Elsevier; 1997. p. 277–91.
[405] Körschens M. Effect of different management systems on carbon and nitrogen dynamics of various soils. In: Lal R, Kimble JM, Follett RF, Stewart BA, editors. Management of Carbon Sequestration in Soil, Advances in Soil Science. Boca Raton, FL: CRC Press; 1998. p. 297–304.
[406] Lal R, editor. Soil Quality and Agricultural Sustainability; 1999.
[407] Leifeld J, Kogel-Knabner I. Soil organic matter fractions as early indicators for carbon stock changes under different land-use? Geoderma 2005;124:143–55.
[408] Otterman J. Baring high albedo soils by overgrazing: a hypothesized desertification mechanism. Science 1974;186:531–3.
[409] Otterman J. Anthropogenic impact on the albedo of the earth. Climatic Change 1977;1:137–55.
[410] Paré T, Dinel H, Moulin AP, Townley-Smith L. Organic matter quality and structural stability of a Black Chernozemic soil under different manure and tillage practices. Geoderma 1999;91:311–26.
[411] Rogers SL, Burns RG. Changes in aggregate stability, nutrient status, indigenous microbial populations, and seedling emergence, following inoculation of soil with Nostoc muscorum. Biol. Fert. Soils 1994;18:209–15.
[412] Sheley RL, Svejcar TJ, Maxwell BD. A theoretical framework for developing successional weed management strategies on rangeland. Weed Technol. 1996;10: 766–73.
[413] Six J, Elliott ET, Paustian K. Soil macroaggregate turnover and microaggregate formation: a mechanism for C sequestration under no-tillage agriculture. Soil Biol. Biochem. 2000;32:2099–103.
[414] Storie RE. An index for rating the agricultural value of soils. Bulletin - California Agricultural Experiment Station. Berkeley, CA: University of California Agricultural Experiment Station; 1933.
[415] Storie RE. The Storie Index Soil Rating Revised Division of Agricultural Sciences. Berkeley, CA: University of California; 1978.
[416] Tongway DJ. Rangeland Soil Condition Assessment Manual. Canberra, ACT, Australia: CSIRO Division of Wildlife and Ecology; 1994.
[417] Tongway DJ. Monitoring soil productive potential. Environ. Monit. Assess. 1995;37:303–18.

[418] van Lanen HAJ, Reinds GJ, Boersma OH, Bouma J. Impact of soil management systems on soil structure and physical properties in a clay loam soil, and the simulated effects on water deficits, soil aeration and workability. Soil Till. Res. 1992;23:203–20.
[419] van Lanen HAJ, van Diepen CA, Reinds GJ, De Koning GHJ. A comparison of qualitative and quantitative physical land evaluations, using an assessment of the potential for sugar beet growth in the European Community. Soil Use Manage. 1992;8:80–9.
[420] von Lützow M, Leifeld J, Kainz M, Kögel-Knabner I, Munch JC. Indications for soil organic matter quality in soils under different management. Geoderma 2001;105(3–4):243–58.
[421] Wu T, Schoenau JJ, Li F, Qian P, Malhi SS, Shi Y. Effect of tillage and rotation on organic carbon forms of chernozemic soils in Saskatchewan. J. Plant Nutr. Soil Sci. 2003;166:328–35.
[422] Yemefack M, Rossiter DG, Njomgang R. Multi-scale characterization of soil variability within an agricultural landscape mosaic system in southern Cameroon. Geoderma 2005;125:117–43.
[423] Young AJ. Agroforestry for soil conservation. Wallingford, England: CAB International; 1989.
[424] Blum WEH, Santelises AA. A concept of sustainability and reslience based on soil functions: the role of ISSS in promoting sustainable land use. In: Greenland DJ, Szabolics I, editors. Soil Resilience and Sustainable Land Use. Wallingford, England: CAB International; 1994. p. 535–42.
[425] Bouma J. The role of soil science in the land use negotiation process. Soil Use Manage. 2001;17:1–6.
[426] Brown RB, Huddleston JH, Anderson JL, editors.. Madison, WI: ASA-CSSA-SSSA; 2000. Agronomy Series No. 39.
[427] Greenland DJ, Szabolics I. Soil Resilience and Sustainable Land Use, vol. xiv. Wallingford, England: CAB International; 1994.
[428] Körschens M, Weigel A, Schulz E. Turnover of soil organic matter (SOM) and long-term balances – tools for evaluating sustainable productivity of soils. Z. Pflanz. Bodenkunde 1998;161:409–24.
[429] Land and Water Development Division Topsoil Characterization for Sustainable Land Management. Rome: FAO; 1998.
[430] Lemieux G. Fundamentals of Forest Ecosystem Pedogenetics: An Approach to Metastability Through Tellurian Biology, vol. 72. Québec, Canada: Laval University Faculty of Forestry and Geomatics Publication; 1997.
[431] Mitchell DJ, Fullen MA, Trueman IC, Fearnehough W. Sustainability of reclaimed desertified land in Ningxia, China. J. Arid Environ. 1998;39:239–51.
[432] Oldeman LR. editor. Guidelines for general assessment of the status of human-induced soil degradation. Global Assessment of Soil Degradation GLASOD: Working Paper and Preprint 88/4, Wageningen, ISRIC; 1988.
[433] Oldeman LR. The global extent of soil degradation. In: Greenland DJ, Szabolics I, editors. Soil Resilience and Sustainable Land Use. Wallingford, England: CAB International; 1994. p. 99–118.
[434] Oldeman LR, Hakkeling RTA, Sombroek WG. World map of the status of human induced soil degradation: an explanatory note. Global Assessment of Soil Degradation GLASOD, Wageningen; Nairobi, ISRIC; UNEP; 1991.
[435] Richter DD, Markewitz D. Understanding Soil Change: Soil Sustainability over Millennia, Centuries, and Decades. Cambridge & New York: Cambridge University Press; 2001.

[436] Smith ML, Bruhn JN, Andersonm JB. The fungus *Armillaria bulbosa* is among the largest and oldest living organisms. Science 1992;356:428–31.
[437] Soil Resources Inventory Group (1981). Soil Resource Inventories and Development Planning: Proceedings of Workshops at Cornell University 1977-1978 Soil Management Support Services (SMSS) Technical Monograph 1. Washington, D.C., Soil Conservation Service, USDA.
[438] Steiner KG. Causes of Soil Degradation and Development Approaches to Sustainable Soil Management. Leiden: Backhuys Publishers; 1996.
[439] Thwaites RN, Slater BK. Soil-landscape resource assessment for plantations – a conceptual framework towards an explicit multi-scale approach. Forest Ecol. Manage. 2000;138:123–38.
[440] van Diepen CA, Van Keulen H, Wolf J, Berkhout JAA. Land evaluation: from intuition to quantificationStewart BA, editor. Advances in Soil Science, vol. 15. New York: Springer; 1991. p. 139–204.
[441] Wilson GA, Bryant RL. Environmental Management. London: UCL Press; 1997.
[442] Zhu A-X. Measuring uncertainty in class assignment for natural resource maps under fuzzy logic. Photogramm. Eng. Rem. S. 1997;63:1195–202.
[443] Zhu A-X. Mapping soil landscape as spatial continua: the neural network approach. Water Resour. Res. 2000;36:663–77.
[444] Zinck JA. Physiography & Soils: ITC Lecture Notes SOL.41. Enschede, The Netherlands: ITC; 1988.
[445] Schiffer MB. Formation Processes of the Archaeological Record. Albuquerque: University of New Mexico Press; 1987.
[446] Gray J, Shear W. Early life on land. Am. Sci. 1992;80:444–56.
[447] Markewitz D. Soil without life? Nature 1997;389:435.
[448] Mojzsis SJ, Arrhenius G, McKeegan KD, Harrison TM, Nutman AP, Friend CRL. Evidence for life on Earth before 3,800 million years ago. Nature 1996;384:55–9.
[449] Navarro-Gonzales R, Rainey FA, Molina P, Bagaley DR, Hollen BJ, Rosa J d l, Small AM, Quinn RC, Grunthaner FJ, Caceres L, Gomez-Silva B, McKay CP. Mars-like soils in the Atacama Desert, Chile, and the dry limit of microbial life. Science 2003;302:1018–21.
[450] Poulet F, Bibring J-P, Mustard JF, Gendrin A, Mangold N, Langevin Y, Arvidson RE, Gondet B, Gomez C.the Omega Team Phyllosilicates on Mars and implications for early Martian climate. Nature 2005;438:623–7.
[451] Stixrude L, Peacor DR. First-principles study of illite-smectite and implications for clay mineral systems. Nature 2002;420:165–8.
[452] Allen CD. Biogeomorphology and biological soil crusts: a symbiotic research relationship. Geomorphologie 2010;4:347–58.
[453] Wardle DA, Bardgett RD, Klironomos JN, Setala H, van der Putten WH, Wall D. Ecological linkages between aboveground and belowground biota. Science 2004;304:1629–33.
[454] Usher MB, Sier ARJ, Hornung M, Millard P. Understanding biological diversity in soil: the UK's Soil Biodiversity Research Programme. Appl. Soil Ecol. 2006;33:101–13.
[455] Zhou Z, Shangguan Z. Dynamic changes of soil ecological factors in Ziwuling secondary forest area under human disturbance. Chinese J. Appl. Ecol. 2005;16(9):1586–90.
[456] Ingham ER. (n. d.) The Soil Biology Primer: chapter 1: The Soil Food Web. USDA, NRCS. Available at: http://soils.usda.gov/sqi/concepts/soil_biology/soil_food_web.html.
[457] Caldwell BA. Effects of invasive Scotch Broom on soil properties in a Pacific coastal prairie soil. Appl. Soil Ecol. 2006;32:149–52.

[458] Heneghan L, Fateme F, Umek L, Grady K, Fagen K, Workman M. The invasive shrub European buckthorn (*Thamnus cathartica*, L) alters soil properties in Midwestern U. S. woodlands. Appl. Soil Ecol. 2006;32:142–8.

[459] Allen CD. Monitoring environmental impact in the Upper Sonoran Lifestyle: a new tool for rapid ecological assessment. Environ. Manage. 2009;43:346–56.

[460] Caravaca F, Alguacil MM, Azcon R, Roldan A. Formation of stable aggregates in rhizosphere soil of Juniperus oxycedrus: effect of AM fungi and organic amendments. Appl. Soil Ecol. 2006;33:30–8.

[461] Jackson W. Farming in nature's image: natural systems agriculture. In: Kimbrell A, editor. The Fatal Harvest Reader: The Tragedy of Industrial Agriculture. Washington, D. C: Island Press; 2002.

[462] Howard A. An Agricultural Testament. London: Oxford University Press; 1940.

[463] Aims KM. Intensification of food production on the Northwest Coast and elsewhere. In: Deur D, Turner NJ, editors. Keeping it Living. Seattle, WA: University of Washington Press; 2005. p. 67–100.

[464] Sorenson R, Bukholm KM, Wu W, Pilo L. Analysis of carbon-fractions in Black Earths from excavations at the Viking-age town Kaupang, southeastern Norway. In: Boschian G, editor. Proceedings of the second international conference on soils and archaeology. Pisa: Dipartimento di Scienze Archeologiche, Universita di Pisa; 2003. p. 124.

[465] Deur D. Tending the garden, making the soil: Northwest Coast estuarine gardens as engineered environments. In: Deur D, Turner NJ, editors. Keeping It Living. Seattle, WA: University of Washington Press; 2005. p. 269–327.

[466] Vitousek PM, Ladefoged TN, Kirch v, Hartshorn AS, Gravew MW, Hotchkiss SC, Tuljapurkar S, Chandwick OA. Soils, Agriculture, and Society in Precontact Hawaii. Science 2004;304:1665–9.

[467] Phillips JD. Stability implications of the state factor model of soils as a nonlinear dynamical system. Geoderma 1993;58:1–15.

[468] Phillips JD. On the relations between complex systems and the factorial model of soil formation (with discussion). Geoderma 1998;86:1–42.

[469] Strahler AN. Dynamic basis of geomorphology. Geol. Soc. Am. Bull. 1952;63:923–37.

[470] Rhoads BL. The dynamic basis of geomorphology re-envisioned. An. Assoc. Am. Geogr. 2006;96(1):14–30.

[471] Huggett RJ. Discussion of the paper by J. D. Phillips. Geoderma 1998;86:23–42.

[472] Jenny H. Derivation of state factor equations of soils and ecosystems. Soil Sci. Soc. Am. Proc. 1961;25:385–8.

[473] Johnson DL, Domier JE, Johnson DN. Reflections on the nature of soil and its biomantle. An. Assoc. Am. Geogr. 2005;95(1):11–31.

[474] Johnson DL, Hole FD. Soil Formation Theory: A Summary of its Principal Impacts on Geography, Geomorphology, Soil-Geomorphology, Quaternary Geology and Paleopedology. Madison, WI: Soil Science Society of America; 1994.

[475] Johnson DL, Keller EA, Rockwell TK. Dynamic pedogenesis: new views on some key soil concepts, and a model for interpreting quaternary Soils. Quaternary Res. 1990;33:306–19.

[476] Young FJ, Hammer RD. Soil-landform relationships on a loess-mantled upland landscape in Missouri. Soil Sci. Soc. Am. J. 2000;64(4):1443–54.